How to Understand Quantum Mechanics

How to Understand Quantum Mechanics

John P Ralston
Department of Physics & Astronomy, The University of Kansas, USA

Morgan & Claypool Publishers

ISBN 978-1-6817-4226-7 (ebook)
ISBN 978-1-6817-4162-8 (print)
ISBN 978-1-6817-4098-0 (mobi)

DOI 10.1088/978-1-6817-4226-7

Version: 20180401

IOP Concise Physics
ISSN 2053-2571 (online)
ISSN 2054-7307 (print)

A Morgan & Claypool publication as part of IOP Concise Physics
Published by Morgan & Claypool Publishers, 1210 Fifth Avenue, Suite 250, San Rafael, CA, 94901, USA

IOP Publishing, Temple Circus, Temple Way, Bristol BS1 6HG, UK

Avoid credit card debt.

Contents

Preface

Sensible people skip the prefaces of books, because they can. When I was young and ignoring bunk, I skipped all of 'em. If you're sensible and also reading this, you must want to know the author's intentions.

Many years ago, I realized that almost everyone was getting more false information about fundamental physics than true information. I am not talking about technology: technology is in good shape. I am referring to finding reliable information to understand physics, so you might advance further, and possibly participate in creating physics. Solid information about basics is very rare. What poses as information about fundamentals is material written by reporters. Reporters think that if a sentence is grammatically correct and uses physics words in a plausible order, then it must be the unique truth about nature. They have never heard of a science where words strung together and making sense could possibly be dead wrong. They would never believe it is possible to combine correct physics equations with exact mathematics and create impossible nonsense. Physics wins all the awards for the *Queen of the Sciences Most Easy to Get Wrong.*

When a person starts learning physics with independence, he or she discovers an unexpectedly high density of distinctly different wrong concepts in arbitrary wrong orders clamoring for attention. People who don't learn with independence don't see it. They freely insert wrong concepts in wrong orders that look OK, because they're not really thinking about it anyway. A very timid and conservative disinformation system then replicates lore about physics the way a cell replicates junk DNA. Meanwhile the priceless original means of discovery get edited out because they are too simple to possibly be important. It really works that way: easy half-page breakthroughs of 50 years ago got recopied into exponentially overblown interpolations by people who just hate easy half-page breakthroughs. Somebody must benefit from it.

This book represents the best way I have discovered to help people understand quantum mechanics. Every step is easy: no steps are left out. We don't make artificial puzzles, and we're not in physics boot-camp. The main secret is to interpret equations just for what they appear to say.

An old tradition dominates. The old tradition is the only part of physics based on bullying people and terrifying them with superstition *not to use or believe the equations for what they say.* It's also loaded with dirty tricks of disinformation. It would be a dirty trick to teach chemistry by saying 'the correct and established *phlogiston theory of fire* can sometimes be calculated using the oxidation aspect of fire. Since the human brain cannot understand fire, we must say that it is both and neither phlogiston and oxidation at the same time'.

The rest is about critically analyzing the terribly illogical ordering of topics and claims of the older approaches. It was a self-serving attempt to salvage the dignity and historical relevance of the old quantum theory by presenting it as pre-requisite for the new theory that destroyed it. How would that make sense? None of the juggernauts who participated in the old quantum theory still survives. Their disciples have disappeared. Our approach being rid of it can cause nobody an insult, except for *the astounding insult caused to the conceptual mistakes of junk DNA still remaining in the system.*

For decades students read out of books whose prefaces emphasized that 'nobody understands quantum mechanics'. Somewhere in this book I observe that's a pretty repulsive way to get started. There's a much better way to put it: nobody understands nonsense. A healthy brain is not supposed to understand nonsense. For decades the healthy minds that rejected superstition had no alternative but to 'shut up and calculate'. In this part of the 21st Century it is finally legal to understand the Schrödinger equation and its solutions, learn everything possible about *entanglement*, insist on descriptions that have no unexplained or supernatural elements, and understand quantum mechanics on its own basis. You can call it an 'interpretation' but it is actually a 'presentation' that is natural, unpretentious, direct, and efficient. That's how physics is supposed to be.

I don't understand professors who say that teaching physics is difficult. I think that teaching physics, and finding out new things about physics, are just about the most fun a person can have. I'm very grateful for students who have been so good as to allow me to do it. Sometimes they ask for 'the most important single fact about the world'. They are surprised when I have the answer right away:

avoid credit card debt

Author biography

John P Ralston

John P Ralston, is a Professor of Physics and Astronomy at The University of Kansas. He is a high energy theorist with more than 30 years experience and publishing on quarks and gluons, neutrinos, leptons and other things wrongly called particles, cosmic rays, cosmology, speculative ideas, and so on. He grew up in Northern Nevada, where physicists were scarce, and knows he's unique in having worked for a living as a material expediter, warehouseman, union electrician and a design engineer. He once re-keyed 200 apartment locks to make survival money, and he's fixed more than one 19th century pocket watch for the pleasure of making it go. He can deal with data like an experimentalist, has won awards for million-dollar research, and has generated licensed intellectual property dealing with real-world applications. Yet for some reason his passion in life is to understand nature in its most simple perfect form, and teach anyone who wants to hear about it.

John P Ralston

Chapter 1

The continuum Universe

Figure 1.1. A cubic mile of space, or a view of the electron wave in an atom, depending on scale. Photo by permission of Tommy Richardsen.

1.0.1 A cubic meter of space

Imagine a cubic meter of space sitting on your kitchen table. It is inside a glass box. The air has been pumped out so you can experiment with space. Space is not a trivial subject. Several physical theories have been proposed for it. The theory that the space you are considering is 'empty' failed a long time ago. It did not explain the data.

Very important data comes from electromagnetism. A large magnet placed on one side of the box will affect a compass needle on the other side. Magnetic fields go right across the box and fill its entire interior, a *continuum*. When playing with magnets and iron filings, one will see a thready, clumpy pattern that appears to be 'field lines'. Actually the magnetic field is smooth and continuous, while the clumpy-thready patterns are the magnetized bits of iron attracting themselves.

doi:10.1088/978-1-6817-4226-7ch1

A concept called 'action at a distance' had proposed instantaneous propagation for magnetic fields (and other fields), but that failed long ago. If you move the magnet, there is a time delay for the fields to cross the box, at a known speed. The changing magnetic fields come with changing electric fields moving inside the box. An impulsive disturbance of the fields travels at the speed of light. Now please! The fields do not fly across the box like little point-like photon torpedoes. One does *not* become more modern or sophisticated about physics by interpreting electrodynamics in those terms, one goes into *mistakes*. The disturbance travels through the field like a breeze-borne ripple in a field of wheat. Except that wheat is too grainy, being a grain food. There is no granularity at all in the continuum: it is perfectly smooth all the way down to zero, as far as the current experiments and theory say. This is the fundamental physical point of view of quantum mechanics. It is about a *continuum*, which probably contradicts what you've heard about photons, and quantum mechanics.

Remove the magnet, turn off the lights, and experiment with the hypothesis of empty space. The hypothesis fails, because the box is full of thermal radiation. Cool the box to the temperature of intergalactic space. The box is still full of colder thermal radiation, your own chunk of non-cosmic microwave background. The wavelengths formally average about one centimeter, but you'd be hard pressed to find a single recognizable 'cosine' shape. The waves occur in a huge range of sizes, shapes, and fluctuations bouncing round in chaotic disorder. *THIS* is quantum mechanical.

There is a theory that black body radiation consists of photons. The theory has been misrepresented and misunderstood. Before 1900, the theory focused on the electromagnetic fields, which are continuous entities sloshing around everywhere inside the box. This was the correct physical picture and a very good theory. The waves of thermal radiation are very jumbled, disorganized, constantly rearranging and shifting in shape, an interpenetrating seething *continuum*. There has never been a time or place empty of them. Beside that, the box is full of many other things more recently found, including the chiral condensate, the Higgs field, gravitational fields, and so on. The theory says they are all infinitely smooth and continuous, and explains data very well.

The misrepresented photon theory says radiation is a bunch of little point-like particles flying around randomly in 'empty space'. This is a perfectly logical interpretation of the words used by physicists, but it is wrong. It is actually impossible to learn physics by being logical and using Webster's dictionary[1] for word meanings: there are too many *internally consistent and logical misinterpretations* which happen to be *wrong*. Moreover, for reasons of history quantum physics became loaded with word abuse, and competition to confuse the physical picture. Coursework concerned with making calculations does not tend to correct it. For that reason, many TV-physics personalities and college physics teachers who passed graduate-level exams never revised the wrong picture

[1] We thank Dr Sherman Sweeney for many observations on this fact.

coming from the misrepresented photon theory of little point-like particles flying around.

The wrong physical picture of the old quantum theory has two errors in the hypothesis of *empty space*, and *little point-like particles.* The actual quantum theory discovered those things do not exist. Quantum mechanics is not about the most tiny subatomic particles. It is not about Planck's constant. Quantum mechanics is about the *Universe*, and the Universe is *big*. Everyone agrees on this. For some reason the facts have a hard time getting out there.

1.0.2 The downside of successful advertising

Figure 1.2. Just as you would guess, a 3D-printing template to make Planck-constant cookie-cutters does exist!. Source: Planck Constant—cookie cutter by cb1986ster, published on 19 November 2015, www.thingiverse.com/thing:1143630.

Winston Churchill wrote that 'History is written by the victors'. Quantum physics is the only part of physics where historical mistakes *before there was a theory* were kept around as disinformation *after* there was a theory. This may surprise you.

We are talking about the *old quantum theory*, which was work done in 1900–25 by Bohr, Einstein, Sommerfeld, Wilson, and many others. The old quantum theory is the domain of formulas like $E = h\nu = \hbar\omega$, and $p = h/\lambda$, or $\vec{p} = \hbar\vec{k}$. The Bohr model was a centerpiece of the old quantum theory (*OQT*), along with the concept of *intrinsic quantization* of energy, angular momentum, and other things. If an argument depends on the finite value of Planck's constant h, or mentions the 'quantum of action', the basis is the *OQT*.

The premises of the *OQT* were that Newtonian point-particles indisputably existed, but needed some constraints to act right in the micro-world. The ideas and formulas of the *OQT* were mass-produced and universally distributed before people knew they were fundamentally wrong. The advertising campaign was successful beyond any measure. It is still around today, maintained by adding a chapter to the storybook. It says the *OQT* and all its basic accomplishments were permanent foundation points, which the new theory made more mathematical, more difficult to understand, while not actually changing the foundations.

The reality was different. Schrödinger's wave theory did not validate the old theory. Instead it exposed the old theory as cookie-cutter equations true *sometimes* when conditions made them circular, but otherwise *false in general.* As a foundation the *OQT* was completely off-track and wrong in every single element! You need this information, because volumes of dis-information will rebroadcast exactly the brain viruses of the *OQT* you don't want in your personal operating system, *which makes understanding so much more difficult.*

Isn't that interesting? Quantum mechanics *now* is not the subject it was believed to be in 1930 or 1950. The name 'quantum mechanics' is a misnomer: the subject is not about 'quantization'. *Physics evolves:* the understanding changes, and the meaning of words changes, while the words themselves stay the same. For many years the presentation of quantum mechanics came from the 1920s, updated periodically to be more technically demanding. The way thinking evolved along the way was secondary to making bigger and more difficult calculations. Like everyone else, the author bought into every element of the *OQT*, and thought it was awesome necessary progress. When he was very young, and not sure of his skills, he had said 'I'm not sure if I am able, but I'd like to understand quantum mechanics *before I die*'. After learning quantum mechanics, and not dying from the experience[2], and thinking about this and other things for more than 40 years, there is a definite conclusion. The *OQT* was not just a dead end. It was a brain-virus dead-end that is bad for you!

Consider this: Bohr, Einstein, Heisenberg, Sommerfeld, and many others had 'quantization', Planck's constant, $E = mc^2$, $E = hf$, $p = h/\lambda$ plus all the skills of top theoretical physicists. Yet all of them were completely defeated for 25 years from 1901–26. If the top brains in physics were stopped for 25 years by the equations released to the public, they will stop you. *The cookie-cutter equations were what stopped progress on the subject. Relying on them in any form will prevent your progress.* This is actually a very helpful discovery. For one thing, Bohr never realized it.

1.0.3 The wrong use of $\vec{x}$

Before quantum mechanics, the symbol $\vec{x}$ meant the position of a point particle, represented by a set of numbers[3] $\vec{x}(t) = (x(t), y(t), z(t))$, for each time t. In one dimension, the motion of a particle under uniform acceleration a is described by

$$x(t) = x_0 + v_0 t + at^2/2.$$

[2] Young physics students tend to be most serious about what they understand the least.

[3] We apologize on behalf of students everywhere that letters x, $\mathbf{x}$, and $\vec{x}$ can all get confused. Some authors use symbol $\vec{r}$, which has its own problems.

Except for loose and colloquial language, quantum physics never uses such a concept of a particle nor a particle trajectory $\vec{x}(t)$. The words 'particle tracks' in a 'particle detector' have changed to words meaning 'quantum field tracks in a quantum field detector'. We only bring it up so it's clear what we are *not* discussing $\vec{x}_N$, with the N having a Newtonian meaning.

This 'point' (sorry for the pun) will become a major theme. If you study and teach quantum mechanics for, say 20 years, you will discover people who make it extremely hard for themselves. Those who fail have the same problem: *they were told and accepted that point particles are on an equal footing with waves.* They imagine $\vec{x}_N(t)$ everywhere but it is never written[4].

It is not an accident. For a long time physics writers had fun interpreting everything calculated for a wave as if it came from a contradictory weird particle. *Quantum tunneling* is an example. The first panel of figure 1.3 shows a wave in one dimension sloshing through (not over) a certain region, emerging on the other side. It is unremarkable: can a sound wave go through a sheetrock wall and be heard on the other side? *Why not!* Yet those who made physics difficult—let's call them the enemies of understanding—have two or three favorite devices. They never consider a movie of the wave oscillating in time, and never show more than one panel from a movie. To help correct that we show several snapshots in figure 1.3. The actual movie is just beautiful!

However, a notion that 'time did not exist in the micro-world' came largely from the Bohr model, which had 'stationary states' where nothing ever happened. When it was discovered to be dead wrong, the zombie came back with presentations removing time

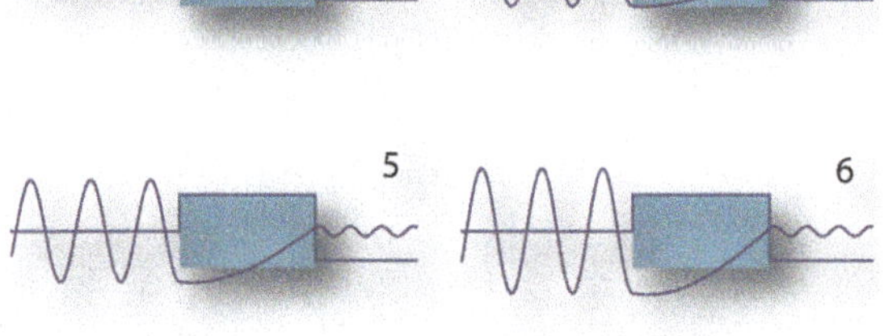

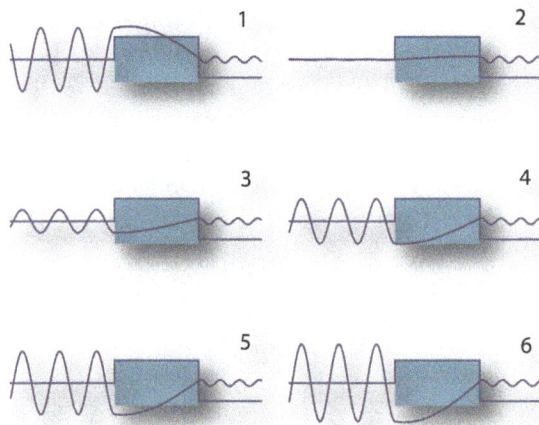

Figure 1.3. Quantum waves are always moving. These movie frames in time ordering show a quantum wave passing through an interaction layer, usually called 'quantum tunneling'. The wave is actually in three dimensions, and the representation is very schematic. It is falsely presented in terms of fictitious particles recycled from the old theory that became irrelevant. The false presentation creates a non-existent mystery.

[4] When you actually study Feynman's path integral, rather than consult physics pop-culture, you will discover it is *not* about particle trajectories.

dependence from quantum mechanics *for purposes of presentation*. Check some sources to discover this: one page of a book has time-dependent factors, the next page strips them off, and then time might disappear forever. Most of the barriers to learning have the same feature: the *OQT* is used to interpret the new theory that contradicts it. Finding waves that are bouncing and jiggling with time-dependence, a little mathematical surgery removes the time-dependence, rediscovering (!) 'stationary states'[5].

Experts don't seem to notice the number of dirty tricks. For example, the colored region in figure 1.3 is reported to be a 'wall a particle cannot cross'. It is not an impenetrable wall, and no calculations nor equations refer to a *parcle*[6]. Anyone looking at the cartoons sees no *parcles* in them, but *the particles imagined to exist from the previous theory that failed* are put in by a juggle of words. Experts don't recognize that the words can do damage. 'The figure shows a particle approaching a barrier'. (It does not.) 'A classical particle cannot cross the barrier. The quantum particle can tunnel through the barrier, to appear on the other side.'

The term 'quantum wave', which is accurate, was deliberately and systematically replaced by another word we hate to write. It was done early while everyone underestimated the impact of the Schrödinger equation when it first appeared. It was at first *unbelievable* that the entire *OQT* had been destroyed by one simple equation. The early presentation based everything on *parcles* with the early intention to keep the old quantum theory intact *forever*, while hoping the Schrödinger wave equation would just go away.

Figure 1.4. Erwin Schrödinger.

[5] For example, Townsend writes that 'the state just picks up an overall phase as time progresses; thus the physical state of the system does not change with time. We often call such an energy eigenstate a *stationary state* to emphasise this lack of time dependence'. The sentences confuse an overall constant phase with time dependent factors.

[6] Since the word *particle* defeats progress, we've made a new word *parcle* for the thing that is not there.

1.0.4 The Enemies of Understanding

Stuffy, stinky old physics books once gave the impression quantum mechanics was so mathematical and sublime that 'the founders themselves' could not understand what it was about. We think that's unhelpful, and an advance form of discouragement coming from an authoritarian mindset. As Feynman once said:[1]

> 'I don't believe in the idea that there are a few peculiar people capable of understanding math, and the rest of the world is normal. Math is a human discovery, and it's no more complicated than humans can understand. I had a calculus book once that said, 'What one fool can do, another can'. What we've been able to work out about nature may look abstract and threatening to someone who hasn't studied it, but it was fools who did it, and in the next generation, all the fools will understand it. There's a tendency to pomposity in all this, to make it all deep and profound.'

Feynman was better than all of them, and he was not pompous. Why are people pompous? We enjoy telling students that physics is like a theater. Most of the players are delightful creative characters in love with the Universe. Here and there the theater has just a few bothersome villains who play dirty tricks. They write 1200 page Big Color books called *Some Kind of Physics* and contribute just the right amount of mistakes to *Wikipedia* to make it disabling. Right? The pompous stuff seems to come from competition to get attention. It's true the founders of quantum mechanics could not always agree with each other. What do you expect? Remember *they were struggling to establish and maintain their careers*, in a very competitive system. Physicists are often jealous, and selfish: they are humans!

Depending on your source, you may find a rather dogmatic presentation that pretends to be weird and wonderful, but it's just dogmatic. Learning quantum mechanics asks you to participate in critical thinking more than most forms of physics. This is because *there's so much bunk in the system. Most of it is word abuse.* If you ask anyone, 'why do YOU think particle physics is about particles?', they may not understand the question. Those 'particle tracks' are not fundamental microscopic phenomena. They are a *macroscopic* chain of correlations you absolutely cannot distinguish from the passage of a fast little wave. Erwin Schrödinger wrote

> 'The cloud chamber and emulsion phenomena, though they are at the moment in the focus of interest, represent after all only a small section of all that we know about nature. In their apparent simplicity they appeal to the vivid imagination of an intelligent child …. Yet they are not as simple as they look. This is witnessed by the pages and pages of intricate formalism that is often devoted to account for even the simplest of them.'

Schrödinger often assumed too much expertise from his readers. He might have mentioned those cloud chamber tracks were the *one and only, sole remaining ghost of the parcle idea*, while also not mandating its use. Schrödinger might have told you that the intricate formalism uses *waves in every calculation, always waves that had been colloquially renamed particles*, because the parcle word was so handy to keep around. And it appears that Schrödinger was unaware of two 1929 papers by C G Darwin and N F Mott that *explained particle tracks as wave phenomena*, eliminating the last vestige of any need for *parcles*. (We'll return to this in chapter 10.)

Physicists are sometimes lazy and complacent with word usage. They can trick themselves and also get tricked. Once the author was at an international meeting where a very angry condensed matter physicist gave a flaming talk that none of the 'particles' of particle physics existed! Nothing claimed to be a new particle lasted long enough to make a track! It was a *great outrage* to perpetrate the hoax of new particles! The complaint was *perfectly correct* about word abuse but also *75 years in the past* in its physical conceptions. The word 'particle' has not in physics meant a dimensionless point for 75 years! That is why Wolfgang Pauli wrote (in a letter to Einstein[7] that the appearance of a point-like electron at a sharp position was 'a creation outside of the laws of nature'.

The pattern of replacing calculations by verbal patterns that don't match is bizarre. Someone must have benefited. Here is an analogy. After thousands of years people finally discovered equations for astronomy that actually describe planetary motion. The discovery that planets are rocks should have killed the theory that planets are demons, but it did not. There were too many people making a living off demons, and the demon business never stopped being profitable. Nowadays celestial mechanics is used to make high precision horoscopes for superstitious rich people, who pay money to believe planets are demons. The Horoscopic interpretation of celestial mechanics uses demon-assumptions to interpret whatever a rock planet might be doing in demon-words. The fact no demon appears anywhere in any equation, while so vividly imagined in the interpretation, is held to be proof that demons are very tricky. Every planet has a demon aspect, and a rock aspect, totally complementary, so when you see one aspect you cannot see the other. The greatest wisdom of the Horoscopic interpretation is that planets are both and neither rocks and demons at the same time. No macroscopic human brain can fully comprehend it.

The term 'Copenhagen interpretation' appeared in the early 1950s. It refers to Bohr's *re-presentation* of the subject after it had been discovered by others. The presentation has heavy emphasis on 'wave–particle duality' and the 'uncertainty principle'. These items had played no role at all in the discovery of quantum mechanics. They play no role in a modern presentation. A person[8] can go to at least 100 physics conferences for 37 years and never hear the term 'wave–particle duality'. No working physicist anywhere views the uncertainty relation as anything more than an easy math fact. However, the core of the Copenhagen *presentation* puts the items early and *in a particular order to make them appear essential*. The presentation of every topic computed for waves was redundantly rehashed by inventing a *parcle*

[7] 'eine ausserhalb der Naturgesetze stehende Schöpfung', as related by H D Zeh 2003 *Decoherence and the Appearance of a Classical World in Quantum Theory* (Berlin: Springer).

[8] Testimonial available on request.

Figure 1.5. There are no demons!.

aspect to be 'complementary'. That is so confusing, and so easy to avoid. How to avoid it? *Dump the particle.*

The order of presentation is everything. Students will ask, 'what difference does the order or presentation make, so long as we see all the material?' It matters because physics is not math. The meaning and usage of physics concepts downstream of one piece of information depends on the ordering upstream. The ordering and decisions of the Copenhagen presentation is the source of physics pop-culture mistakes that quantum mechanics is full of paradoxes and cannot be understood. That quantum physics would *never be understood* was Bohr's life-long philosophy, also adopted by Heisenberg before there was a quantum theory, from which the two never deviated after the actual theory arrived. When Schrödinger's theory first appeared and *was absolutely easy* to understand, Heisenberg dismissed it as 'crap' for that very reason.

One should be repelled by a presentation advertising[9] it was not going to be understood. We say everyone in the 21st Century *must* understand quantum

[9]The preface of Griffith's book cites Bohr for saying 'If you are not confused about quantum mechanics, then you have not really understood it'. There's a related quote from Feynman, but surely he was joking!

mechanics. That calls for re-ordering the material and avoiding the tradition of listing 'axioms of quantum theory', posing as if physics were high school Euclidean geometry. In fact quantum mechanics is the only subject in physics where teachers traditionally present haywire axioms they don't really believe, and regularly violate in research. (Most violations occur with the fishy *eigenvalue postulate*, see section 8.3.) The schoolbook subject also tends to be authoritarian about obeying rules without understanding the reasons, a result of rules put out of order that do not follow logically. The logical disorder and incipient contradictions of an outmoded presentation can be avoided, and that makes learning quantum mechanics so much easier.

Some claim that physics is based on principles. But basing decisions on principles not understood makes a person weaker, not stronger. If a person understands the basis of information, he or she needs no authority nor their principles to justify it. So it is pretty well known that people cannot learn physics when it's based on authority. That's why US students cannot learn physics from high school football coaches[10]. *Memorizing authority statements that are not understood is why people can't understand quantum mechanics.*

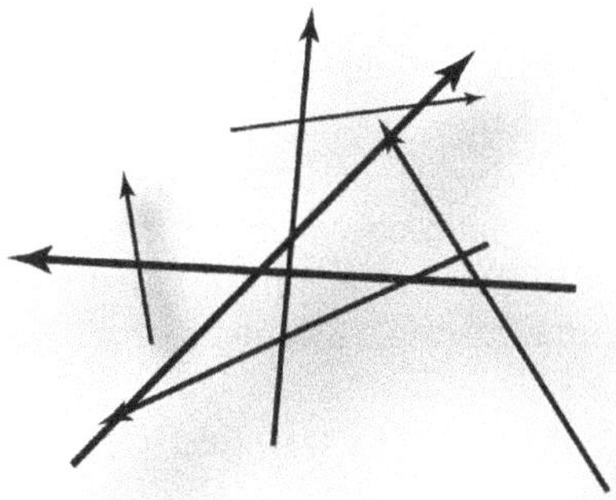

Figure 1.6. Don't look at this picture! It shows what people with the wrong concept of $\vec{x}$ are thinking about.

1.1 The right use of $\vec{x}$: describing a continuum

Water is a great example of a continuous substance. It is compressible (although very strong), it supports arbitrary wave shapes in its interior, and it definitely exists. Water has no special shape, and simply conforms to the shape of the forces acting on it. Predicting the future of a pail of water is *really challenging* because so many details are involved. Water in the space shuttle forms itself into bouncing spherical blobs. That is typical wave physics. It's a good prototype for the electron wave trapped in an atom, except the water blob has a rather sharp boundary electron waves seldom have. Also, first fill the space shuttle with the substance *everywhere*.

[10] Football coaches should not be offended. The 1200 page Big Color Books are so bad that US high school students can only learn physics from super-inspired teachers, who are rare.

The electron wave is like a pulse or a lump of stressed water inside the water. There's no nothingness outside the electron!

To describe a continuum mathematically, you first propose (or declare) that the underlying space, formerly considered 'empty', provides a continuous coordinate system with labels $\vec{x} = (x, y, z)$, or similar. The default is that no points are themselves actually different than others, and that the labels do not themselves represent anything physical. Choose the vicinity of some arbitrary point for the discussion. In that region there is some disturbance, a function $\psi(\vec{x}, t)$. The physical variable *at* symbol $\vec{x}$ is the wave $\psi_x(t)$. We changed notation from $\psi(\vec{x}, t) \to \psi_x(t)$, which is read 'psi-sub-x as a function of *t*'. The first thing you notice is that there are as many time-dependent physical variables as there are points in space. (Later we'll discover there are quite a few more, for more species of fields.)

This physical picture is just the opposite of the point particle. The disturbance represented by the wave function is 'everywhere it happens to be'. One can usually localize the 'big region' of the electron wave pretty well: we can say that an atomic electron wave is spread over the volume of an atom, which is pretty small on the human scale. Yet outside the atom the electron wave does not just disappear. The amplitude of the wave approaches zero: zero is a definite value, not the same as disappearing.

The point is not semantic, but about consistency. The wave function is (are) the time-dependent dynamical variable(s), or 'degrees of freedom'. Each point in space has a dynamically independent degree of freedom, coupled by the equations to other, nearby points. All those dynamical variables are expressed by a very big *list* $\psi_{\vec{x}}(t)$ for what happens at each arbitrary point $\vec{x}$. Everywhere the disturbance is small, we may dismiss it as uninteresting, but the physical variables don't disappear. If a wave, namely the physical variable *at* a particular position $\vec{x}= (1, 2, 3)$ is zero, the physical variable $\psi((1, 2, 3)) = 0$ has not been disturbed from zero, and still waits to participate in physics.

The Schrödinger equation predicts how a given quantum matter wave will change in time. 'Time evolution' means a wave shape $\psi(\vec{x}, t = 0) \to \psi(\vec{x}, t = t_1)$, which is a new shape at time t_1. The resulting shape is a given shape, and time evolves by $\psi(\vec{x}, t_1) \to \psi(\vec{x}, t_2)$, to make another new shape; see figure 1.7. The Schrödinger equation predicts the time rate of change by

$$\imath\frac{\partial\psi}{\partial t} = \Omega\psi, \tag{1.1}$$

where Ω will be called the *frequency operator*. The notation Ω is innovative and self-explanatory, since $\partial/\partial t$ has dimensions of *frequency*.

Operators are math tools that might have been used in Newtonian physics. The equation $\vec{v} = d\vec{x}/dt$ predicts velocity by applying the time derivative operator d/dt to $\vec{x}$. We could invent an operator that acts on $\vec{x}$ and returns $\vec{v}_{system}(\vec{x})$ for a given *system* Then classifying and memorizing interesting velocity operators, we would build up the description of systems one by one by solving $d\vec{x}/dt = \vec{v}_{system}(\vec{x})$. (That plus theorems on existence and uniqueness of solutions is how a course in differential equations works.) We could invent an operator that acts on $\vec{x}$ and returns

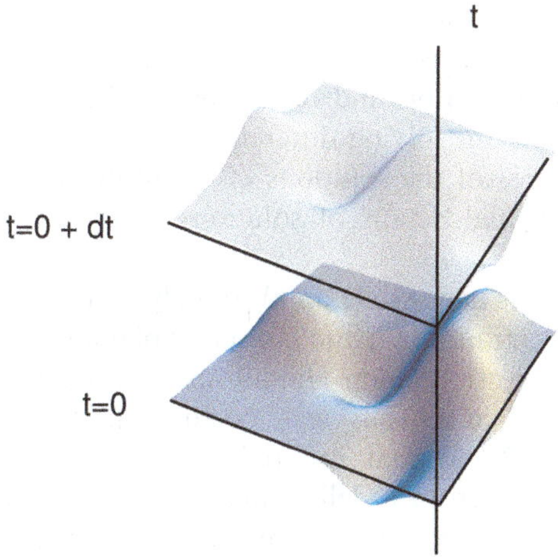

Figure 1.7. A graphic with time moving 'up' for how wave equations predict time evolution. No particular 'shape' of a function is actually determined by the equation. The initial shape is entirely set by initial conditions. The equation uses shape information from values and curvatures at each moment in time t to predict the function's new shape at updated time $t + dt$. The shape 'time evolves', and the wave moves.

'acceleration' $\vec{A}_{system}(\vec{x})$ for each system. The corresponding differential equation could be $d^2\vec{x}/dt^2 = \vec{A}_{system}(\vec{x})$.

Instead of 'velocity operator' we prefer 'frequency operator' in $\imath\partial\psi/\partial t = \Omega\psi$ because of the $\imath$ in the equation. A deceptively easy example is $\Omega = \omega_0$, some constant. The differential equation predicts $\psi(t)$ by

$$\imath\frac{\partial\psi}{\partial t} = \omega_0\psi; \quad \rightarrow \psi(t) = \psi(0)e^{-\imath\omega_0 t}. \tag{1.2}$$

This is harmonic time dependence with angular frequency ω_0. There is a subtle difference between representing a system by a differential equation and simply giving a solution to the system's time dependence. The differential equation does *not* depend on initial conditions and does not predict initial conditions. This represents our guess that nature defines the system, which will be the same system for any initial conditions. Removing initial conditions is the accomplishment of an equation of motion, and it is the only accomplishment. That's because the set of all solutions, for all initial conditions, is precisely equivalent to a differential equation, and vice verse. Did you know that? The differential equation is a concise *code* for all solutions with all initial conditions *left unspecified* and nothing more.

No differential equation as easy as equation (1.2) will ever describe a wave. Waves have infinitely variable shapes, and then infinitely many initial conditions. We'll review this using example systems that are not too difficult. If it's going to predict realistic waves, system by system, a realistic operator Ω must involve some serious math. We will postpone the details of Ω for now. Be warned that in the traditional approach the most difficult cases were presented early as difficult puzzles. That's something of a professor-cheat and also obsolete. Before 1930, say, physics was largely based on solving differential equations, which usually meant memorizing the solutions to the small class that could be solved. After 1930, and with the organizing principles of operators and vector spaces of quantum mechanics, physics was much less dependent on solving differential equations. It is very important to know the *general features* of the solutions of quantum mechanics, which are very easy. Meanwhile the actual details of solutions become less and less important: unless you need them.

Our use of symbol Ω is new, while an absolutely equivalent symbol H called the Hamiltonian operator is traditional and *not* self-explanatory[11]. The road forks here. One road leads to a digression about Hamiltonians, certain historical prescriptions for what was 'allowed' and what was postulated, and often some dogmatic assertions (see section 7.1.3) that are redundant. Along that road were little analogies with *parcles*, often at a deliberately high math level students could not challenge. After a big loop all roads come to *time evolution*, which is our road. Quantum mechanics is mostly about time evolution, which is *causal and deterministic*, as is clear from the Schrödinger equation of motion. Since we are concerned with causal, mechanistic time evolution, it would be inappropriate to confuse the analysis with references to probability until they are needed. That is how quantum mechanics is done. First you characterize the wave function, and solve what it is doing. *After* the calculations are made, statistical predictions are extracted: not during the process!

1.1.1 The wave function describes the state

Notice that the word 'function' has subtly changed its meaning. In math classwork a function is a map from an input to an output. Students score points by evaluating functions presented as a puzzle and computing the right output. That meaning is not being used, and we discourage you from evaluating wave functions, except when using a computer to make graphics. It is better to consider the wave function as being pre-evaluated by a computer graphics routine, or an equivalent wave-*list* that represents the state of the system.

The concept of a well-defined state began with Schrödinger's wave-*list* ψ. There was no correct concept of a state in quantum physics before it. The attempt of the Bohr model to define a 'state' was so atrocious that in 1925 Born, Heisenberg, and Jordan set up a theory called 'matrix mechanics' where the concept of a state did not appear. The competition between matrix mechanics and Schrödinger's wave theory led to disinformation reporting they were 'the same' theory. They are not the same: matrix mechanics had no concept of a quantum state, and no place to put the initial conditions of a quantum state.

Within a year of discovering his equation, Schrödinger was able to derive all the correct features of matrix mechanics as a consequence of it. The converse is impossible, and without a concept of a state matrix mechanics did not predict

[11] When equation (1.2) is multiplied on both sides by $\hbar$, the equation is unchanged, and $\hbar$ cancels out. Defining $H = \hbar\Omega$ produces $i\hbar\partial\psi/\partial t = H\psi$, where $\hbar$ constant still cancels out.

quantum mechanics. We're mentioning this because matrix mechanics has symbols '$\vec{x}(t)$' and '$\vec{p}(t)$' that stand for time-dependent *operators*. These symbols are often introduced with equations looking much like Newtonian ones, which is a notational trick that causes great confusion. It is premature to say much about such operators, other than warn you the symbols *do not describe parcle* trajectories. They are mathematical tools that nicely automate a few easy calculations, make all other calculations impossible, and play no role in the basic understanding of quantum theory. One can understand it with a single paragraph in section 9.1.4.

With the modern picture of the state of the system as a little bouncing wave, what seemed to be a particle moving around was an unresolved blob of *disturbances* getting swapped from one location to another. Whether or not you buy this, that is how the math is set up and it is how the math works. The dynamical variables labeled by fixed points don't fly around in space. The *time-dependent deformation* propagates among the cheerful little wave variables, like a continuous crowd performing 'the wave' at a football stadium. Some may think it is more advanced to replace water with educated facts about water molecules, and the stadium wave by the individual people. It is not helpful here. The smooth and continuous behavior of water, which you know by experience and see with your eyes, gets to the soul of the Universe. Also try to remember the molecules themselves are not gritty little particles, but continuous interpenetrating jiggling quantum *waves*.

Later we'll talk about your thumb, which from the quantum point of view is quite a remarkable thing. For now, imagine someone with poor eyesight seeing a bowling ball under a bed sheet. When the ball rolls around they might think it's a ghost flying atop the sheet. Actually the bed sheet moves up and down to create the illusion: the *disturbance* under the sheet moves from place to place. If the ball is taken away, the bed sheet goes back to its zero-deformation state: it does not disappear. The wave function is like the bed sheet: we are aware of its movable disturbances, and call it 'matter'.

While quantum mechanics needs many variables to describe waves, the dynamics itself is rather plain and easy; see figure 1.8. Many people have the relations all mixed up, with notions of 'operators' and 'probability' competing and conflicting

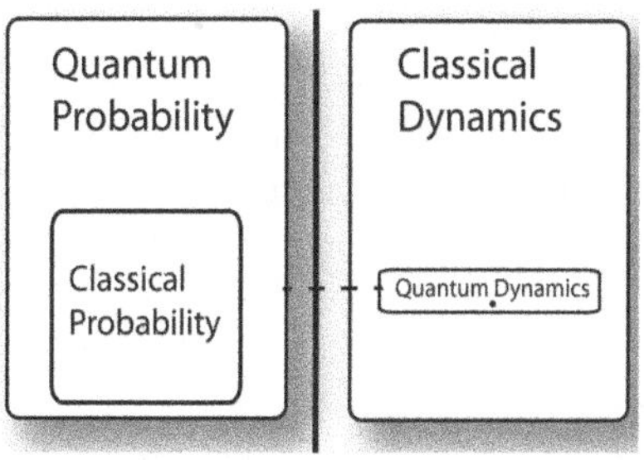

Figure 1.8. The big picture of quantum theory. Quantum *dynamics* is a very special linear type (the dot) inside of the larger scheme of classical Hamiltonian and Lagrangian dynamics of *wave systems. Newtonian* particle dynamics includes non-linear cases, but uses far fewer variables. Quantum probability is more or less separate and disconnected; the links between the two are tenuous.

interpretations. As figure 1.8 suggests, we can develop almost all of quantum theory in terms of basic *wave dynamics* that absolutely agrees with human intuition (about waves, of course!)

In a quantum mechanical universe, a cosmic ray proton kicked out of some distant supernova does not become a Newtonian or relativistic particle traveling across light years of empty space. Those particle ideas did not work out. They do not exist now, so we can't use them. The cosmic ray proton is a *resonant excitation* of the quantum mechanical stuff which propagates itself as a self-sustaining disturbance in the ground state of the Universe. This quantum mechanical ground state is really big, and probably a new concept to you. Being a completely new idea, it is given an old and maximally misleading name: 'the vacuum'.

1.1.2 FIAQ

Here are frequently and infrequently asked questions, *FIAQ*:

There's something different about this presentation. Aren't we supposed to start with Planck's constant? Maybe if we lived a hundred years ago. *Absolutely no*, for the 21st Century. If you have an $\hbar$ tattoo we recommend you have it removed[12]. The consternation that eliminating $\hbar$ causes to some is amazing. But since the *OQT* was addicted to $\hbar$, you can and should avoid its substance dependence.

Consider this: the concept of a little wave does not, by itself and on its own merits, need a conversion factor of the year 1900 into *MKS* units defined by *Newtonian* physics. Atomic spectra were first measured as *frequencies,* and continue to be measured as *frequencies.* If the Schrödinger equation can calculate frequencies directly—and it does!—then that suffices to 'do physics'. The intermediate step of the *OQT* involving the Newtonian energy of *particles* was a digression. The mix-ups[13] over $\hbar$ are so many that we'll need to return to this. For now know his: there are no exceptions to the rule that reference to $\hbar$ sends you backwards into ancient history. Dispensing with $\hbar$ takes you forward.

There's something different about this presentation. Aren't we supposed to start with the probability to find a particle? Why? And NO! Since we don't need point *parcles* any more, and will never use them, then why define physics using what does not exist?

Unless you come from Mars, or have a *very lucky* educational background, the *one and only* defining fact of the wave function $\psi(\vec{x})$ previously seen will be a claim[14] that 'the probability to find the particle in the volume $dxdyz$ is $\psi^*(\vec{x})\psi(\vec{x})dxdydz$'. This one 'postulate' plants four independent *mistakes.* It defines quantum probability in terms of distributions, which later is contradicted by the *Born rule* (see section 9.1.6.) that does not refer to distributions. It instructs people to replace complex $\psi(\vec{x})$, loaded with information, with $\psi^*(\vec{x})\psi(\vec{x})$ that has no useful computational information. It suggests you can use the claim freely in intermediate

[12] The symbol $\hbar$ is pronounced 'h-bar', which led to more than one nerdy establishment serving beer to wanna-be theorists. The experimentalists drink at the 'Error Bar'.

[13] Gaffes, gaps, scams, blunders, bungles, boo-boos, boners.

[14] The textbook by Liboff is particularly rigid about this mistake.

calculations, which is false. And it initiates an illogical structure that must have a *parcle* for its existence, which is wrong.

Physics evolved. Bohr's 1929-ish Copenhagen presentation came and went. It is true that the wave function is a humanly constructed proxy for nature, and not nature itself. *What else do you expect?* The probabilistic nature of quantum mechanics can't be dumbed down, and the naive definitions found in basic books include *mistakes* about probability. Let's not start there! To learn quantum mechanics, you *absolutely should* believe you are faithfully describing a little wave, because that will teach you how it works. Later, your little wave calculation is converted to *probabilities for little waves.* Physa-bloggers and TV experts carelessly using the words quantum *parcle* are the enemies of understanding!

How can we possibly think about waves without a particle? Don't we need particles to make waves? No, there's been some brainwashing about needing particles as thinking tools. You have never seen a particle of air nor a particle of water. Your human experience with those funny continuous substances is priceless. It's a bad idea to replace your valuable intuition by an imaginary 19th Century gas of little jelly beans. In every experiment so far the Universe has been 100% continuum, contradicting the *partcle* conceptions, which must be dropped. There are no quantized orbits, no intrinsic quantizations, no quantum leaps, no clouds of point-like electrons, no irreducible disturbances of measurements. All that collapsed when Schrödinger discovered quantum waves.

Shouldn't the credit for quantum waves go to de Broglie? Partly: we can be generous. History is also a form of power. We recommend fact-checking original sources. In his 1924 dissertation de Broglie wrote:[15]

> 'Considërons d'abord le mouvement rectiligne et une forme d'un mobile libre. Les hypothèses faitesau début du chapitre premier nous ont permis, grâce au principe de Relativité restreinte, l' étude complète de ce cas .... Ici nous devons poser:
>
> $$\begin{gathered} \nu = \frac{W}{h} = \frac{m_0 c^2}{h\sqrt{1 - \beta^2}}, \\ \sum_i p_i dq_i = \frac{m_0 \beta^2 c^2}{h\sqrt{1 - \beta^2}} dt = \frac{m_0 \beta c}{h\sqrt{1 - \beta^2}} d\ell = \nu \frac{d\ell}{V}, \end{gathered} \tag{1.3}$$
>
> d'ou $V = c/\beta$. Nous avons donné une interprétation de ce résultat au point de vue de l'space-temps.'

Consider the problem from two perspectives, if any. At the minimum, hundreds of millions of people have uncritically seen the equations, with their teachers saying they were meaningful. That is because those people were told the discussion was about particles, and they did not know any better.

[15] de Broglie L 1924 Recherches sur la théorie des quanta *PhD Thesis* Université de Paris, France.

From the other perspective, think again: if you propose a wave, what are those things *mass m, velocity V*, and momentum *p* that appear in the equations? They are particle concepts in the context of de Broglie and Bohr. Those masses and momenta are not wave concepts nor wave attributes, in the context of the original. That is why de Broglie's equations never made sense as a theory nor a piece of a theory, when viewed on their own. The problem for de Broglie, then and now, was attempting to assemble cookie-cutter equations obtained on one basis into a new concept that needed new equations. Upon hearing about it, Peter Debye ridiculed the method, saying, 'if this man is serious, why does he not write a competent wave equation?' (And Schrödinger followed up on it.)

The topic called 'wave–particle duality' amounts to 99.9% commitment to particles, and 0.1% vaguely associating a cookie-cutter wave to make the particle misbehave. The attempt is not internally consistent, and not a part of quantum mechanics. In 1924 de Broglie's idea was far-seeing. In this century it is a bunch of symbols seeking their meaning in the mistakes it kept on board: you are better off not translating them from French to English. There are many examples of how random combinations of equations might seem to predict new physical information, but do not.

Writers often present physics history as going in a straight line, with every praised accomplishment leading directly to the next. A sort of phony reverence for 'the founders' can be a cover for a sequence actually based on accepting authority statements without critical thinking. The actual progress of physics research, and physics history, is as 'crooked as a dog's hind leg'. Twenty five years were wasted between Planck's correct *empirical fit* to the black body spectrum and Schrödinger writing a viable wave equation. There are *no cases* where the pre-quantum equations became foundations for quantum mechanics. Every pre-quantum equation derived from data was found to be a special case, and not a general principle. Watch out for those who order their presentation chronologically, and apply patches to make it look coherent: it falls apart when you study the actual subject. You will encounter false advertising, and false information on what to believe. *Let's not* repeat a full century of physics by building on *mistakes*.

References

[1] Feynman R P 2001 *The Pleasure of Finding Things Out: The Best Short Works of Richard P. Feynman* (London: Penguin)

[2] Schrödinger E 1952 *Br. J. Philos. Sci* **3** 233

[3] Zeh H D 2003 *Decoherence and the Appearance of a Classical World in Quantum Theory* (Berlin: Springer)

[4] de Broglie L 1924 Recherches sur la théorie des quanta *PhD Thesis* Université de Paris, France.

IOP Concise Physics

How to Understand Quantum Mechanics

John P Ralston

Chapter 2

Everything is a wave

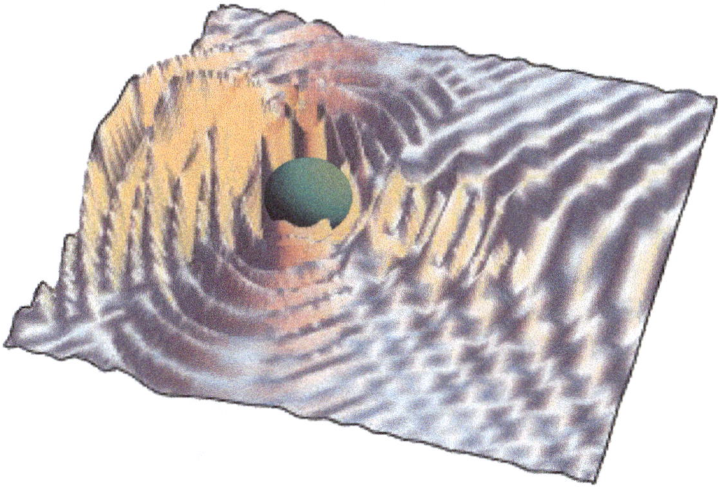

Figure 2.1. A picture of a quantum wave scattering off an impenetrable sphere. This comes from an exact calculation, not an artist's conception. The wave moves from left to right in oscillatory fashion: a movie would be ideal. The real part of the wave function is shown as the height; the actual wave exists in three dimensions, and is rotationally symmetric about the axis of propagation. The imaginary part is simply 90° out of phase. The reflection is the pile-up on the left. The 'shadow' downstream on the right is actually a nebulous phenomenon, not a sharp one. There are no edges, and color was assigned by arbitrary software. *Almost all* quantum waves are *almost always* moving.

Everything known—solid, liquid, vapor, or plasma, the vacuum, out past the furthest galaxies, plus the space in between—is some kind of wave. Folks, the quantum universe is just *beautiful.* You cannot know the gorgeous, sensuous, voluptuous, infinite dimensional beauty until you understand quantum mechanics.

doi:10.1088/978-1-6817-4226-7ch2

Everyone in physics reports that learning quantum mechanics was the most staggering experience of their life.[1]

When people hear this news, they are not sure they believe their ears, or what the the word means. *Waves in what?*.

2.1 Waves in what medium: waves made of what stuff?

Let's add a bit to the picture. We are not talking about waves like those on the surface of water. Such waves occupy a boundary where the water ends. Quantum waves exist throughout the *interior* of the Universe, where there are no boundaries.

To get started, humans are very good at understanding a lower-dimensional world, like a two-dimensional surface. It is a first step to developing and trusting your understanding. Then consider a mental movie of an unremarkable wave moving and spreading over the surface of a distant ocean. If your movie is just ordinary quality—not approaching a well-made TV commercial advertising water—it might have more informational content than learning volumes on partial differential equations. It may seem a paradox that even a child's vision of waves could have so much information. There is no paradox. Mathematics is often a clumsy under-representation of how people think. It is possibly less powerful than how the insects can think, but it has still developed into the most powerful tool to *communicate and assist* our thinking.

Once comfortable with a two-dimensional wave, gently back away from using the 'height' of a wave in centimeters. There will be a 'value' of the wave, also called its *amplitude*, which is more general. The 'amplitude' will be expressed in units to be decided later: be flexible.

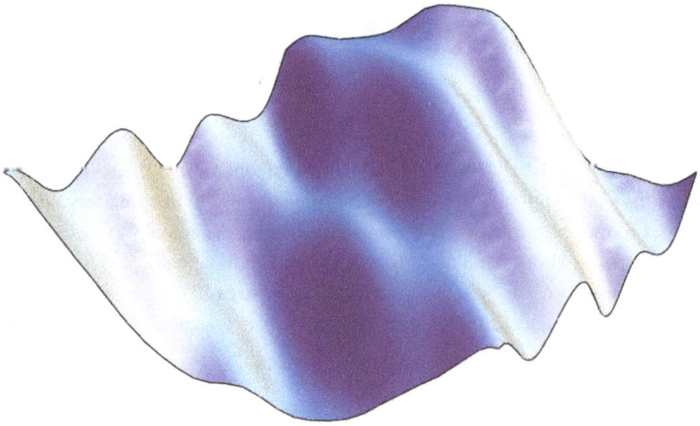

Figure 2.2. A bird's eye view of a generic electron, which is an ever-vibrating wave with no particular shape. This is actually a picture on a two-dimensional space. The real thing lives in three dimensions. There are no edges.

[1] Some say nerds need to 'get a life in the real world'. Which real world?

Now consider waves in three-dimensional space. Imagine what it means in the context that 'the world is made of waves'. The waves must be all around us, inside us, interpenetrating. In fact they are ceaselessly vibrating, intermingling, inside of everything and one another. For example, the proton is a little self-trapped wave, like a smaller atom inside the electron wave that makes an atom. In a convenient approximation the proton is ignored, and imagined fixed, but that is an approximation. The electromagnetic field waves pass right through and interpenetrate both the electron and proton waves, and vice versa. This is the new and simple vision. Don't bother with Mr Mach's particular hobby, a picture of empty space with a moving point. Start thinking of the entire Universe as living inside a continuous multi-dimensional ocean without a boundary.

Then what is the stuff of waves? It is the stuff of what the Universe is!

2.2 Evidence for waves

Visible light waves are rather small. Very short waves tend to go in straight lines, known as 'rays'. The straightness comes from the *symmetries* of physics in three-dimensional space. The direction and wavelength of the wave are conserved, namely do not change with time.

This is another example of the relation between symmetries and conservation laws. In fact it is the explanation of *conservation of momentum*, referring to the momentum of *light rays*, not particles. No particular place is special in a vacuum. Once moving in a vacuum, there's no special place to stop: waves must keep moving, so wave momentum is conserved.

It is not very hard to deduce that light is a wave, and Mr Huygens, the contemporary of Mr Newton, explained a great deal of it. It is sad that wave–particle apologists tried to rewrite history, which was always clear on the point that waves could explain everything observed, while Mr Newton's particles never did. Somewhat like Huygens, the author's mother first explained the rainbow of color from an oil slick on a puddle of water. 'Look at that!,' she said. 'You always see those colors. Oil is not colored. What causes that? The color is the light not the oil. It is caused by the oil layer being very thin, and stopping waves. The colors not stopped

Figure 2.3. A modest house in the neighborhood, seen through a pinhole. Small distortions from light waves diffracting off the pinhole become easier to see if the pinhole is wobbled slightly with the line of sight fixed.

reflect back, separated and pretty.' My mother disgraced Mr Newton, who botched the interpretation.

'Stealth airplanes' became public when an American president accidentally mentioned they existed while he was on TV. The press was later invited to see the technology, supposedly explained by radar-absorbing paint. We believe in good paint, but there's probably more to the trickery.[2] Nobody notices that the front lenses of binoculars were being stealth-coated for visible light 50 years ago. The coating has layers of material to cancel reflections. When reflections are nullified, more light goes into the binoculars to make a brighter image. It's not increased absorption, it is increased *transmission*.

2.2.1 Do this experiment

With small effort you can play with waves of light. Make a clean pinhole in paper with a pin or a sharp pen or pencil. A 0.5 mm mechanical pencil works well. Aluminum foil or a gum wrapper works better than paper, if any is available. Look through the pinhole at something detailed and far away outside on a sunny day. Without moving your eye, wobble the pinhole very slightly, and try to observe some distortion near the edges of the image. A slight watery distortion should be visible, as if looking through a somewhat flawed lens. The experiment is not critical, and you can even make a pinhole with the cracks between your fingers. Wobbling the pinhole helps your brain resolve small angular deflections, and it eliminates the question of whether the position of your eye is to blame.

Human eyes can resolve angular features of 1/1000–1/10 000 rad. That's equivalent to seeing a spot 10 cm to 1 cm across (grapefruit to gummy-bear size) at a distance of 100 m. Light waves going through a pinhole diffract sideways by a definite angle. It is approximately the wavelength divided by the diameter of the hole. Since you see the effect, the wavelength of light is somewhere around 1/1000 of the pinhole size. Dividing 1/2 mm by 1/1000 gives a wavelength of 1/2000mm. This is dead on. Ignore the accident of being dead on. The wavelength of light *cannot be arbitrarily far* from the estimate. You are now an experimental physicist.[3] You can check the particular wavelength is good for oil slicks, the optical coating of binoculars, designing space telescopes, and so on.

2.2.2 La tache de Poisson, also called Arago's spot

Light emerging from a pinhole fans out like ocean waves passing a rock. Downstream from a rock the waves coalesce as if the rock were never there; see figure 2.4. François Arago (1786–1853) used the same effect to test the wave theory of light. He sent light through a pinhole[4] and past a tiny metal disk stuck on a piece of glass. (Tiny metal disks are very easy to make. Melt a small bit of metal into a tiny

[2] The thing to suspect is active elements, like noise-canceling headphones.

[3] We're not perfectly sure the watery distortions are 100% due to the wave disturbance, because the finite size of the eye's pupil is a complication. The experiment, like all, shows *consistency* of a theory, data, and an interpretation.

[4] The pinhole isolates a pure and simple source. Without a pinhole source, or a diverging lens, light might trivially get behind an object by coming in sideways.

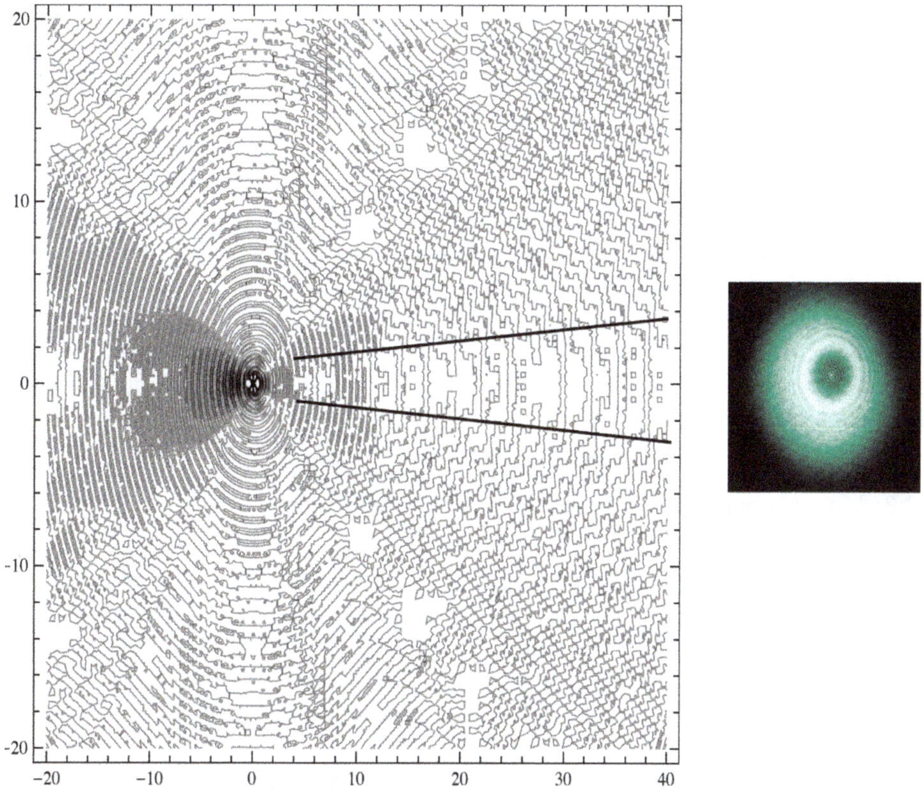

Figure 2.4. Two views of a bright spot downstream and inside a shadow. The left panel shows a calculation of waves moving left to right. The dark lines show the approximate shadow boundary, inside of which waves will merge and reappear. The right panel is from an experiment using a laser.

ball. Tap the ball with a hammer.) The light emerging from the pinhole clearly diverges, and the shadow of the disk was observed to grow larger with distance. Far downstream, however, a bright spot appeared in the center of the shadow: the 'spot of Arago' (or 'la tache de Poisson'). Light did not go through the metal disk, but curved *around the disk*, and the coalescence downstream proved light is a wave. Careful observation can even reveal bright and dark rings inside the shadow. The experiment is both qualitative and quantitative. The quality of light being a wave was tested and confirmed.[5]

The distance down the optical path where a central spot would be significant was also predicted. This is not trivial, and involves the physics of diverging rays (Fresnel zone) many advanced physics books do not even cover.[6] The furthest distance to see

[5] We don't actually believe in science of confirmation. Interpreted fairly, the idea that was tested happened not to fail.

[6] Fresnel made calculations with spherical waves, which are more realistic than 'plane waves' of formally infinite extent. To second order in approximation the phase of spherical waves has a term going like the square-root of the distance times the object size, then divided by the wavelength. Such terms are lost in plane-wave approximations.

the spot is near the area of the disk divided by the wavelength. For Arago's reported 2 mm disk and wavelength 1/2000mm, the furthest distance is 25 000 mm, or 25 m. That's too far for a table-top experiment in 1818, but with a disk one third the size, a ninth of the distance is 2.6 m, which looks good.

Imagine the accomplishment in 1818. Waves of the nearly infinitesimal length of 1/2000 mm actually curve in free space. Arago had made a microscope to look at light itself.[7] But Arago had help. Augustin-Jean Fresnel had developed a wave theory that explained the strength of reflection and refraction from glass or crystal surfaces. Simeon Poisson's theory held that light was a stream of particles. Poisson attempted to shoot down Fresnel's waves on the basis that a bright spot would be observed far downstream of shadows. Poisson thought that was absurd, and proposed the test. Arago decided to test the absurd. Poisson attempted to falsify the wave hypothesis, but ended up *truthifying* it.

2.2.3 Most photon waves are much larger than most atoms

There are many subtleties in cookie-cutter equations, and here is one of interest. A 60 Hz AC circuit should have a wavelength $\lambda \sim c/(60\ \mathrm{s}^{-1}) = 5000$ km. On that naive basis, every wire in every town of the American electrical grid should be exactly in phase with every other. But that's too simple. If you investigate the industrial zone of any town, you'll see banks of capacitors the engineers have strategically located to get the phase of the local three-phase electrical supply to behave itself under varying loads of electrical usage. An electrical power grid is an immensely complicated electrical wave *medium* where unwanted space and time fluctuations cost real money. The engineers get the wave relations right because their calculations are never distracted by the mistaken concept of 'point-like photons'. Once 60 Hz AC is supplied to its windings, a basic electric motor rotates in synchronization with the applied time-dependence, and the user never thinks about the large-scale variations.

Recall that a typical ray of visible light has a wavelength of 1/2000mm. Since microscopes cannot see atoms, they must be much smaller than a wavelength of light. To be 'steady' a wave must repeat nearly the same for many periods. Suppose $\Delta\nu \ll \nu$ is the range of frequencies. After a number of about $N \gtrsim \nu/\Delta\nu$ waves, the wave train will repeat. The total time for N periods is $\Delta t \sim N/\nu \gtrsim 1/\Delta\nu$. This is called the *uncertainty relation* for time and frequency,

$$\Delta t \Delta\nu \gtrsim 1.$$

It adds detail to the obvious and circular fact that an absolutely periodic signal ($\Delta\nu = 0$) must last forever ($\Delta t \to \infty$). While it was quite oversold, there's actually not much to the uncertainty relation: it is a math identity about Fourier analysis. More information will be given in section 8.2.4.

When atoms make an atomic transition, the frequency spread $\Delta\nu$ is often exceedingly small compared to the central frequency. The observed frequency

[7] Savvy ingenious technicians probably contributed assistance to Mr Arago without attribution, just like happens now.

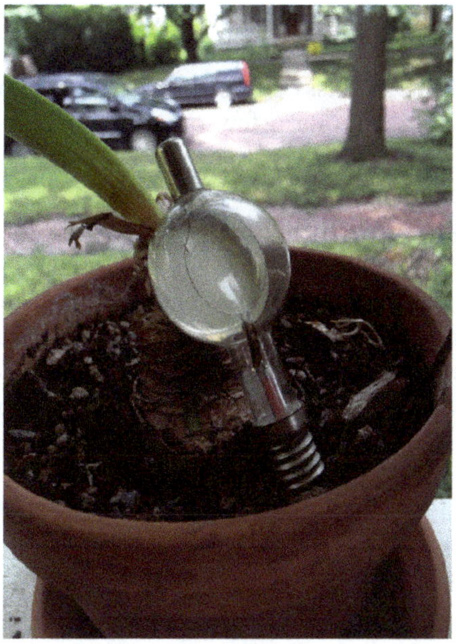

Figure 2.5. A quaint 1930s phototube in a flower pot. Light shining on the metal backing kicks electrons out to the wire loop, which has a voltage set up to attract them. An atom explodes because of a *resonance* between its time dependence and the light.

spread measures the time for the transition to occur: it completely contradicts the 'instantaneous quantum leap' assumed in the Bohr model. The $1S2S$ transition of hydrogen is an extreme example. The intrinsic width $\Delta\nu \sim 1\text{Hz}$, and compared to the transition frequency $\nu/\Delta\nu \sim 10^{15}$. The atom oscillates like a radio antenna for about a second of time while emitting the radiation. Moreover, the transition produces *two* photons, violating the *OQT* rule assuming one photon must be emitted or absorbed at a time. The $1S2P$ transition is faster: it takes a nanosecond. Light travels about 1 ft in a nanosecond, but 1 ft is *hundreds of millions* larger than the atomic size. Both the size and time scale numbers are completely inconsistent with the 'photon' being localized anywhere on the scale of the atom.

The physical situation is so far from the point-like photon picture that it is breathtaking. On the TV show *Star Trek* the photon torpedoes are dangerous pinpoints of light shot out like cannon balls. Nothing like that ever happens. The calculation of an atomic transition is more like watching a plastic beach ball respond to a tsunami wave in mid-ocean. The beach ball is not the photon, it is the electron wave trapped in the atom. The tsunami wave is the photon. A typical mid-ocean tsunami wave may be 200–300 km long, depending on ocean depth. Its velocity may be 1000 km h^{-1}. The numbers mean a mid-ocean tsunami wave is very flat and boring. Try to imagine the scale of flatness 300 km long compared to a beach ball. If all else is glassy smooth—the default for quantum theorists—the beach ball slowly rises and slowly falls, as a long, flat wave takes 20 min passing under its butt.

Two tenuously attached beach balls—a molecule—might fall apart from the interaction. Imagine watching two beach balls struggling over their conjugal destiny over 12 h of gentle sloshing in glassy smooth water where nothing at all seems to be happening. No, please don't. Quite a few quantum processes take a long time to occur, and they are gentle, non-destructive, peaceful unions and disunions nothing like the cannon-ball excitement of TV shows. When and if physicists can arrange for any localized, hard collision, it never comes from the localization of 'particles', which have never been observed.

However the correct physical picture was unavailable in the *OQT*. Einstein, Bohr, and almost everyone misunderstood the effects and importance of *resonance*. The resonance match of *time dependence* was missed due to the dogma of 'energy levels'. When resonance was overlooked the matching of conditions was wrongly thought to be a match of *subatomic spatial collisions* that never happened. Many people have experienced an open metal pipe howling mysteriously in a strong wind. The pipe might be pretty small, but there's no demon in the pipe. The resonant match of time dependence caused by a breeze miles in extent causes the effect. Similarly, the sharpness of the transition observed in the photoelectric effect is due to the resonant response of the atom, as a nanodetector, not photons!

In 1968 Willis Lamb and Marlon Scully surprised many with a paper [2] called 'The photoelectric effect without photons'. The paper simply repeated the quantum mechanical calculations found in textbooks since the 1920s *while not using the language or words of 'photons'.* No trace of a particle-photon or a quantum-photon appears in the calculation. Around 40 years before, people should have started reporting that 'we found out the photoelectric effect comes from a resonant response of electron waves in the atom'. Instead, a bias to confirm and validate the pre-quantum *OQT* had caused the calculations to be consistently presented in words presupposing photons, and presupposing the methods of the *OQT*, which Lamb and Scully noticed *were never used in the calculation.* To this day the 'patches' made for gaps of pre-quantum theory are regularly used for the post-quantum theory that has no gaps.

The question goes to what one expects from a physical theory. A complete theory has no gaps, and never appeals to external postulates for support. Einstein, Bohr, and the others conceiving the *OQT* were not fools, and they were *very visibly aware* that making ad-hoc postulates *contradicted the proper design of a physical theory*. Yet after quantum theory a struggle reappeared to go back to ad-hoc postulates. Physics is supposed to be better than that. Someone[8] has written 'The public is more familiar with bad design than good design. It is, in effect, conditioned to prefer bad design, because that is what it lives with. The new becomes threatening, the old reassuring.'

Edward Tufte, noted for designing beautiful books, wrote that

> 'Good design is clear thinking made visible, bad design is stupidity made visible.'

---

[8] Attributed to Paul Rand.

2.3 Early clues to the size and nature of atoms

In this century you can buy a machine to resolve and manipulate single atoms. Doing physics without hi-tech and extravagant expenditure of money needs real cleverness. To measure the size of atoms with 19th Century technology, here is what to do.

Liquids and solids are basically incompressible: the 19th Century picture has atoms packed tightly like identical balls in a box. For an estimate, an atom can be moved in any single direction by about one atomic diameter. Small molecules are close to the size of atoms, so we'll ignore the distinction for a while.

Nowadays you can find liquid nitrogen in any hospital. Its Newtonian mass density is 0.807 g cm^{-3}, the atomic weight of N_2 is 28.0, so 1 mole of the molecules make 28 g = 28/.807 = 35 cm^3 of volume. The volume per atom is 17.5 $cm^3/6.02 \times 10^{23}$ $=2.8 \times 10^{-23}$ cm^3. That volume will be close to the molecule size-cubed, which gives *size* = 3.3×10^{-8} cm, or *zero point three-three* millionths of a millimeter. Besides assuming cube-shaped molecules, this calculation has a flaw. It relies on Avogadro's number 6.02×10^{23} taught in school, without giving the information on how you would ever determine Avogadro's number. Avogadro himself had no method to find his own number.

And so, for a long time Avogadro's number was not known. There is information in gases. A gas consists of atoms heated to separate and fly around freely. Heating a given volume of liquid or solid in vacuum not too far above the vaporization point makes a gas with about 10 000 times more volume than the corresponding solid. This depends on the temperature, so consider steam at 200 °C ~500 K. Since the cross sectional area of the molecule does not change, the extra volume of gas comes from the volume through which atoms move, called the 'mean free path' (MFP). (You can imagine randomly oriented solid rods about one MFP in length and the diameter of an atom filling most of the space.) The volume fraction predicts the

Figure 2.6. Do not look at this figure. It's from an obsolete government agency with an obsolete logo describing a dysfunctional and bad design that never existed.

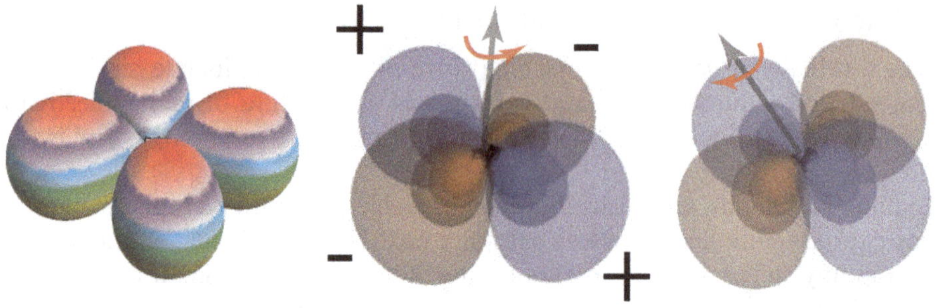

Figure 2.7. Three cartoon of the hydrogen $n = 3$, $\ell = 2$, $m = 2$ wave function. The left 'Easter eggs' picture uses three conceptual errors dating to the 1920s. Mistake #0 omits the time dependence. All the states are oscillating, and some whirl around like propellors. Mistake #1 shows contours without the information there is a continuous three-dimensional structure, like layers of an onion, except a perfectly smooth onion. Mistake #2 is making pictures of the modulus-squared $\psi^*\psi$ at all. It is rather useless. The panels on the right show a few transparent layers of the real part of the wave function. A sense of rotation is indicated by the axis and arrow. The colors stand for *positive and negative regions* of the wave function. The atom on the right happens to have its rotation axis tilted. Such things naturally happen.

MFP is about 10 000 times the atomic diameter. Now we need the MFP from 19th Century data.

The MFP was deduced early from *diffusion*. It describes the slow spread of atoms or molecules that are over-concentrated in a region. An excess number density ρ spreads by current $\vec{j}_D = D\vec{\nabla}\rho$, which defines the *diffusion constant D*, with dimensions $D \sim length^2/time$. For an irresponsible estimate, imagine a bottle of perfume in a closed room is opened very carefully 10 cm from your nose. You might smell it in a few seconds: 10^{-1} s is too fast, and 1000 s is too slow. That estimates $D \sim 1\ \mathrm{cm}^2\ \mathrm{s}^{-1}$. D will be an increasing function of molecular velocities v, and by dimensional analysis $D \sim Lv$. The length scale L must be the MFP, because there's nothing else. From the perfume geruch-experiment

$$\mathrm{MFP} \sim \frac{D}{v} \sim \frac{1\ \mathrm{cm}^2\ s^{-1}}{3 \times 10^4\ \mathrm{cm\ s}^{-1}} \sim 10^{-5}\ \mathrm{cm};$$
$$d \sim 10^{-4}\ \mathrm{MFP} \sim 10^{-9}\ \mathrm{cm}.$$

Calculations by Johann Loschmidt in 1865 found a typical MFP for a gas at room temperature and pressure to be around 10^{-4} cm. Dividing by 10 000 gives molecular diameter: 10^{-8} cm. Improving the estimate with the volume of a spherical atom being $\pi(diameter)^3/6$, and so on, Loschmidt claimed to find an accurate $d = 8 \times 0.000866 \times 0.000140 = 0.000000969$ mm, (his original numbers), which he sensibly reported 'or in round numbers one millionth of a millimeter for the diameter of an air molecule'. Lofshmidt attributed the first estimate to Maxwell and to later work by Meyer.

Maxwell subsequently cited Lofschmidt and an independent 1868 determination using the same method by George Johnstone Stoney (1826–1911). Stoney happens to be quite central to the early development of quantum physics. He did more than anyone we might name. Yet for some reason, Stoney, perhaps for being a brilliant

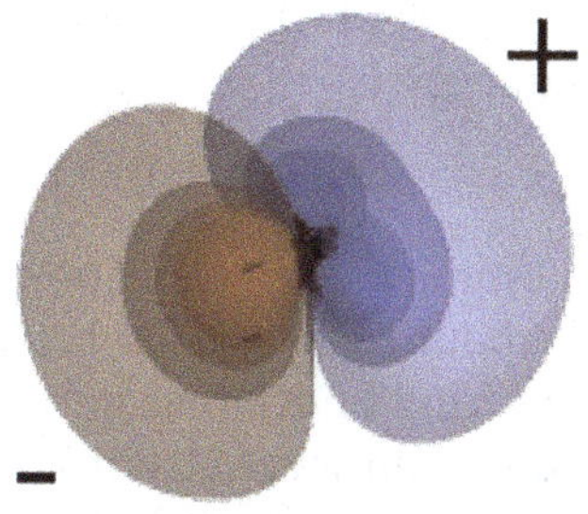

Figure 2.8. A more typical electron wave in a hydrogen atom that has been 'kicked' into a random blobby shape. The colors show the sign of the wave function at the instant of the snapshot.

Irish genius of the 19th Century, was completely ignored[9] by the later period of Planck, Einstein, Sommerfeld, etc, having their turns at being European geniuses. Stoney measured the size of atoms, first deduced the existence of 'an atom of electricity' he called the *electrine*, correctly attributed the frequencies of atomic spectra to the motion of electrines, and fought back successfully when Helmholtz appropriated the atom of electricity and renamed it the 'electron'. Stoney also determined the unit of elementary charge, up to the experimental uncertainties of the time. Knowing the unit of electric charge allows you to electrolyze 18 g of water into 22.4 l each of H and O_2 at *STP*, over-determining Avogadro's number. All this was done by Stoney before 1891. Stoney also set up the pre-quantum atom with elliptical orbits: more on this in section 6.1.2. It was not the ultimately correct direction, but it is the atom you were told was the genius idea of people working 20 and 30 years later. If people had followed up correctly, many troublesome wrong ideas might never have gotten to be so popular.

2.3.1 How to use the size of an atom

The size of an atom is a curiosity before you know atoms are little waves. The facts of waves helps us use the size productively.

As a crude estimate, the size of trapped electron waves is somewhat smaller than the atom. Unless matter waves go faster than light, we expect the physical wave speed to be less than light speed $c = 3 \times 10^8$ cm s^{-1}. To order of magnitude, the frequency of electron wave vibrations $f \sim speed/size \lesssim 10^{-8}$ cm/(10^8 (cm s^{-1}) $\lesssim 10^{16}$ s^{-1}. An atom at rest, and vibrating constantly at a frequency approaching 10^{16} Hz is a breathtaking conception.[10]

Compare the calculation to experiments, remembering the exponent 16. Since they are vibrating, atoms should emit light. Following Mr Balmer, Mr Rydberg's observations were well-described by an emitted-light frequency formula $f = R_\infty c(1/n_1^2 - 1/n_2^2)$. We'll explain the integers n_1, n_2 momentarily. In one

[9] Stoney was famous in his own time, and actually in 1861 elected a Fellow of the Royal Society, a top honor.
[10] The unit of Hertz, abbreviation Hz, has two functions. It is an annoying thing to memorize in place of *cycles per second*, showing how international committees annoy scientists. The Hz unit is also distinct from the radians per second unit of angular frequency, $\omega = 2\pi f$, introducing another 2π to be memorized.

surviving 1888 record, you can find the constant $R_\infty = 1.097 \times 10^7\ \mathrm{m}^{-1}$. The corresponding vibrational frequency is

$$2\pi R_\infty c = \omega_H = 2.08 \times 10^{16}\ \mathrm{Hz}. \tag{2.3}$$

We estimated 10^{16} Hz: as an estimate, the calculation is right on. The Schrödinger equation predicts everything about hydrogen by beautiful, self-contained calculations without any flaky volunteering of external information or estimates.

The frequencies of light a quantum wave absorbs or emits are the frequencies of its electromagnetic current. By a formula from the Schrödinger equation (see section 6.1.4), the current's frequencies are the *differences* of frequencies with which the wave vibrates. Rydberg's formula for light frequencies is the difference of two frequency formulas, just as consistent. The $n_1 = 1$ case shows the lowest possible frequency a hydrogen atom can have[11] is ω_H.

We have just reviewed information that is supposed to be well known. However, the way we are approaching it is new. We have not mentioned or used Planck's constant anywhere. We have not used or mentioned any concept of intrinsic quantization of anything. Those elements of the *OQT* are not just unnecessary, but they simply do not exist along the modern path to quantum mechanics. Those were the mistaken ideas that stopped Bohr and others for decades, so that they could not accept the actual subject after it was discovered.

Let's repeat the physical picture. The atom is a trapped electron wave. The wave does not just sit there like a dot. It is vibrating very rapidly, and it cannot stop, because the balance between wavy springiness and wavy bouncing is what an atom *is*.

2.3.2 The aether came back!

To repeat, the stuff of quantum waves is the Universe itself. There was a precedent in 19th Century aether theories. They generally assumed that empty space contained a mechanical substance which supported the vibrations of light waves. Many have heard that Maxwell and Einstein and relativity eliminated the aether. It once was said 'there are just equations, and no underlying medium is consistent'. That is *true* in physics pop-culture, *false* for physics as it now exists. Pop-culture is always 100 years out of date!

The aether did not go away, but came back as the *Lorentz invariant vacuum.* There had been a concept error associating a dynamical medium with a special Galilean rest frame. No such relation exists, as shown by constructing a *mechanically correlated medium* that is perfectly dynamical, supporting all kinds of waves, yet which has no preferred *relativistic* rest frame. In your first course in quantum field theory (which is somewhat above the level of this material) you will spend weeks developing Lorentz-invariant, classical continuum aether theory (*LICAT*)

[11] The bound state frequency is *negative*. It still represents ordinary vibrations, but with a phase relation that makes waves self-trapping.

Figure 2.9. The dynamical framework of quantum mechanics is continuum 'jello' theory. We live inside the jello.

while never using the word 'aether' or acronym *LICAT*. Please do some fact checking: such material is usually found *after* defining fields, the 'action principle', generalized Euler–Lagrange equations, the facts and representations of the Lorentz group, Noether's theorem, and before 'field quantization'.

In *continuum mechanics* every disturbance of the 'medium' is described by a coordinate q_x and its conjugate momentum p_x. These *never* describe a point-like *parcle*. Instead they refer to continuously varying sets that are infinite by their very nature. You would do the same to describe a wave propagating inside a block of jello. The main difference between quantum mechanics, which uses continuum mechanics, and the Newtonian model lies in a *continuous infinity of dynamical variables*, which cannot be reduced to three 'position' coordinates.

Yet as soon as we mention 'momentum' the brainwashing by Newtonian physics might be misinterpreted as a *parcle* existing point by point in the medium. That is *not* the right idea, but originates in *mistakes* of Newtonian physics incompetently attempting to define everything with a *parcle*. By 1800 Lagrange's formulation of mechanics had revealed that *momentum* is a very general concept, not equivalent to the Newtonian one, however many times it is misunderstood. Fields and waves have momenta, and infinitely many of them, for the infinity or motions they undergo. *All of the mixups about momentum* come from one terrible and wrong 'definition' $p = mv$, and from ignoring the more flexible and universal definition of continuum Lagrangian and Hamiltonian physics. Since we cannot explain that subject in one paragraph, try to remember this: quantum mechanics of waves always has an infinite number of dynamical variables. Every attempt to dumb the variables down to those of a classical Newtonian *parcle* causes mistakes and problems. Every authority who tells you the wave function 'does not exist as a physical wave, since there is no aether for light or matter' is parroting unhelpful advertising that became obsolete. Physics evolved and moved on, while the advertising stayed around.

Ironically, the Michelson–Morely experiment, so much cited to support special relativity, actually did have a preferred rest frame. The Ewald–Oseen extinction theorem [1] takes into account the propagation of light in air and other media. Light does not jump across empty space between molecules, but interacts with constant regeneration of its waves among the molecular waves, and adopting the overall molecular rest frame. The effect is subtle, and directly contradicts a supposed estimate (actually a bluff) that the index of refraction of air (1.000 28) is 'negligible'. After traveling a mere millimeter of distance in air, light has already adapted itself to the overall rest frame of the air. Then an experiment done in air has no information about 'propagation in free space'. The Michelson–Morely experiment was conducted in air and measured the speed of light in different directions of the atmosphere of the laboratory room. Not surprisingly, the motion of the Earth through space was not detected.

The Ewald–Oseen theorem appeared in 1915–16, many years after the Michelson–Morely experiment and special relativity. The Michelson–Morely experiment was published in 1887 and is available online [3]. It appeared 36 years after Fizeau's 1851 water tube experiments *did* observe a change in the speed of light traveling with and against the flow of moving water. Michelson and Morely had repeated and confirmed Fizeau's experiments before building their famous test apparatus. Repeated experiments and repeated theories of thoughtful, dedicated experts did not settle down into a consistent picture for decades due to one overarching fact: *waves can be very, very subtle.*

2.3.3 FIAQ

Many writers say the Schrödinger wave function is just a tool to statistically predict the action of particles. The presentation here claims the wave function exists physically. Which is it? This is a question about 'existence'. The people who told you what exists lied to you. They don't know what exists: we don't know what nature 'is' either. Then why did those writers insist on *parcles*?

Here is an important fact: interpreting wave equations in precisely the way they are written and the way they work is internally consistent, and makes quantum mechanics most easy to learn. Learning physics always involves making choices of what issues to postpone, or completely ignore. Faithful, methodical people who try to memorize everything in a 1200 page book do worse than those who find the actual information and ignore the book.

We concentrate on quantum waves as potentially the literal, real substance of the Universe because it might be true. *It is not automatically wrong.* Believing the equations in front of you is also maximally efficient to learn the material. The most satisfying approach makes interpretations fit the equations, while final decisions about 'ultimate reality' are not needed until they are needed. In that order of development the wave function is as real as anything else.

By the time Fresnel and Arago were investigating waves of light, it was *already known* that the wave concept was not exhaustively complete. The proposal that *classical light* was an electromagnetic wave was not exhaustively complete. Light with a given polarization will pass a polarizer oriented to pass it, and stop on a polarizer at right angles. The orientation of any particular electric field is directly observable. There is *no* electric field which will pass with 50% intensity through every orientation of a polarizer. Unpolarized light cannot be described by a wave amplitude, nor any superpositions of wave amplitudes. Since the electric field model has this flaw even in *classical* physics, is the electric field an artificial mathematical construct with no physical existence?

Matter waves for electrons also have *polarization*. There is a Schrödinger wave model for an electron ignoring that, and a different wave model incorporating it. Just as for light, there is no wave function for an electron with polarization that is also unpolarized. So there is more to learn: quantum theory in general form uses a more subtle tool called the *density matrix*. A density matrix is not always needed, and the way it works is equivalent to a wave function *sometimes*.

The over-selling of quantum mechanics made it hard to get reliable information about it. The first efforts to command and control the subject went off track, setting up postulates in a pretended Euclid-style axiomatic framework. Since that was done out of order, quantum mechanics is the only topic in physics where postulates were set up and contradicted by quantum mechanics. Since the general case disobeys most of the pretended postulates, it is greatly suppressed. The presentation of the density matrix is either omitted, or described as: 'In some cases we do not have complete information about a wave function'. (Mostly we NEVER have complete information about ANYTHING.) A truthful statement would be: 'We generally guess the description by a wave function which might be adequate. It is not always so, and in general, systems are described by density matrices'. Since we've given you this information, we will *not* be making stinky axiomatic postulates about wave functions. You would not accept any!

Qualitatively, a density matrix allows a system to be described by adding up weighted predictions from a number of wave functions. You would do this on your own to compute weighted averages, which the density matrix automates. Strange to tell, the overly restrictive postulates of the *OQT*, as replanted by the Copenhagen presentation asserting the 'eigenvalue postulate' (see section 8.3), strictly says[12] this is 'not allowed'.

To understand quantum probability one must first learn to use *one* wave function with the concepts of *entanglement* (see chapter 9). Coherent presentations beginning with density matrices, and arriving at wave functions for special cases do exist[13]: it is not an easy road. The main advantage is the realization that quantum mechanics is *descriptive*, not *proscriptive*: nothing is prohibited until you make a model with prohibitions. What else did you expect? We suggest a person should return to pondering ultimate reality after learning the basic physics, and discovering *how very many times* people have lied to others about 'quantum reality'.

[12] It's true. Those postulates made to seem so important contradict themselves, so they are ignored regularly.

[13] See, e.g. Peres A 1993 *Quantum Theory: Concepts and Methods* (Dordrecht: Kluwer).

References

[1] Born M and Wolf E 1970 *Principles of Optics* 4th edn (Oxford: Pergamon)

[2] Lamb W and Scully M 1969 The photoelectric effect without photons *Polarization, Matière et Rayonnement,* Volume in Honour of A Kastler (Paris: Presses Universitaires de France)

[3] Michelson A A and Morely E W 1887 On the relative motion of the Earth and the luminiferous ether *Am. J. Sci.* **34** 333

[4] Peres A 1993 *Quantum Theory: Concepts and Methods* (Dordrecht: Kluwer)

Chapter 3

There is no classical theory of matter

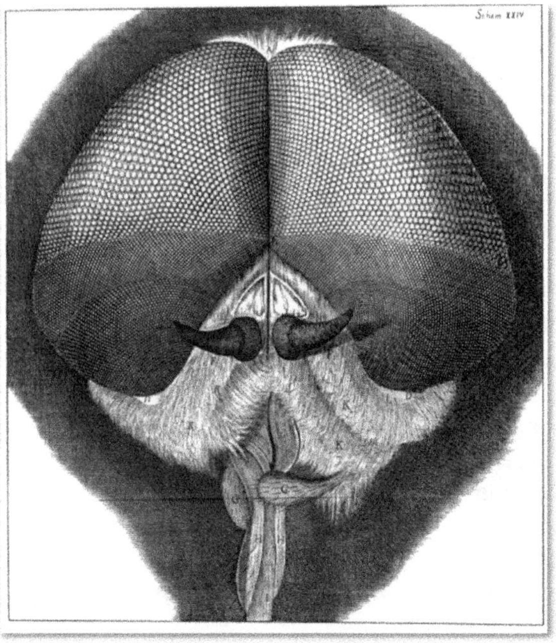

Figure 3.1. Head of a fly, drawn by Robert Hooke and published in his *Micrographia*, 1665. Hooke was an acute observer of nature, and deduced universal gravitation and the conception of the *vector field* of gravity extending across space.

The discovery of universal gravitational attraction by a force decreasing with the inverse square distance was due to Robert Hooke. He published it in his *Cutlerian Lectures* of 1665, and supported it with observations of the Moon, the variation of atmospheric pressure with height, and of course Kepler's laws. That was done more than 20 years before Newton's *Principia*. It was Hooke's challenge to Newton to reproduce the result, and possibly improve it, which first drew Newton's attention to the problem.

doi:10.1088/978-1-6817-4226-7ch3

That is probably not the history you have heard: as Winston Churchill said twice: 'History is written by the victors'. Newton was the victor who hated Hooke, arranged after his death to have every portrait of him burned, forged documents, and removed the substance of the *Cutlerian Lectures* from Hooke's posthumously published collected works[1]. Kepler, Huygens, Hooke, Wren, Euler, and many others made the breakthroughs of the 'clockwork Universe' for which Newton has sometimes been given all the credit. Yet the subject called Newtonian physics never led to a consistent Universe. Classical mechanics cannot deliver a consistent theory of matter. Ultimately this is due to dimensional analysis, which in Newtonian physics has a major flaw called the *kilogram*, and which in electrodynamics has no dimensional scale at all. Quantum mechanics is flawless and contains its own inherent dimensions. It *does* possess fundamental constants with units of length, or frequency, depending on your preference, and it *does not* have room for the Newtonian *kilogram*.

3.1 Earnshaw's no go theorem

Earnshaw's theorem says there's no stable equilibrium configuration for any classical point electric charge. Stable equilibrium means that if the forces on an object are balanced to zero a small variation of the object's position pushes the object back into position, rather than kicking it out of position. The proof is easy. The total external forces summed over a sphere surrounding a given charge must point inward. By Gauss' law the sum of external electric fields over a sphere is proportional to the *external charge inside the sphere*, which is zero at the given point. (The force of the given charge on itself is never counted in classical electricity and magnetism.) 'Magnets' won't help. They have no effect on electric charges at rest. Their effects on each other obey Earnshaw's theorem.

The upshot is that if you arrange a bowl-shaped configuration of potential energy for a classical charge's position in one direction, there must be a saddle-shaped compensation in another direction. If you attempt to stably balance a bowling ball on a horse's saddle, it will fall and hurt someone on one side or the other.

Let's make this quantitative. The electrostatic potential[2] is a classical *field* $\phi(\vec{x}, t)$. Fields generally obey wave equations, in this case *and* for a region without external charges

$$c^2 \vec{\nabla}^2 \phi = \frac{\partial^2 \phi}{\partial t^2}. \tag{3.1}$$

The restriction 'no charges' is more sneaky than you might think. If your theory has point charges, there are no charges *everywhere*, except for infinitesimal mathematical pathologies (singularities) exactly on top of the charges. So the equation applies

[1] Hooke was so badly mauled by history that one needs to check original sources. His complete *Cutlerian Lectures* were published and did survive. If you are interested we recommend beginning there. Two noteworthy papers are [1] and [2]. A possible discovery of a lost portrait is related in [3]. Another source is [4].

[2] Experts recognize the electrostatic potential to be a 'piece' or a 'feature' of the more general electromagnetic vector potential, but that does not affect the argument.

right up to the edge and all the way around infinitesimal points, which is pretty much 'everywhere'.

Before you get equation-hives, let's do something easy. The *static* assumption is zero 'speed' and zero 'acceleration' of the field everywhere, $\partial\phi/\partial t = \partial^2\phi/\partial t^2 = 0$. The right hand side of the equation is zero. Consider a two-dimensional situation $\phi(\vec{x}) = \phi(x, y)$. (This does not mean the world is two-dimensional. It means ϕ does not depend on z, so it's unchanging in the z direction. It's a case to consider.) We then have

$$\vec{\nabla}^2 \phi \rightarrow \frac{\partial^2 \phi}{\partial x^2} + \frac{\partial^2 \phi}{\partial y^2} = 0. \tag{3.2}$$

A bowl-shaped trapping region along the x direction needs increasing curvature $\partial^2\phi/\partial x^2 < 0$. Then $\partial^2\phi/\partial y^2 > 0$ follows for the y direction, which represents an upside-down bowl shape that is unstable. Right?

Earnshaw's theorem is more general than Poisson's two-dimensional electrostatic formula, equation (3.2). One reason to show you Poisson's formula is to observe something unexpected, and not well covered anywhere. The solutions to equation (3.2) and its three-dimensional version *have no definite shape*. You can literally and without restriction 'bend' the potential function into *any shape you desire* along one direction, such as 'x'. The equation responds by computing $\partial^2\phi/\partial x^2$, everywhere you specified it, and demanding payback among all the other terms A function in three-dimensional space has three natural curvatures[3] at every point, which can be taken as $\partial^2\phi/\partial x^2$, $\partial^2\phi/\partial y^2$, and $\partial^2\phi/\partial z^2$. If you've taken calculus—or even if you failed calculus—you know that each of those three symbols is a particular, computable quantity. Poisson's formula does not specify three computable quantities, but only the *sum* of three quantities. Since it's not predicting much, the equation cannot possibly predict any particular shape, and only places one gentle restriction on all possible shapes of the electrostatic field. We wish this basic fact was more widely known!

Just as with the *OQT*, a few cookie-cutter electromagnetic formulas have circulated that give a short-run edge for engineering applications, while doing at least as much harm in the long run. We are thinking about the 'Coulomb solution', an electrostatic potential $\phi = e/|\vec{x}|$, $\phi = e/r$, $\phi = e/(4\pi\varepsilon_0 r)$, in different notations. The formula describes a singular spherically symmetric configuration of pre-ordained shape, directly contradicting the general rule we just discovered, which predicts no special shape. In early work the formula is derived from Coulomb's law, reported to be a law, which seems pretty serious. Then Coulomb's law and all its baggage seem to *predict and require* the point-like *parcle* we said we'd never need. The contradictions are resolved by asking which concepts and formulas are more fundamental than others. We began with the wave equation (3.1). It includes time-dependence, while Coulomb's law does not. Coulomb's law, as written, predicts

[3] A different use of the word 'curvature' involves mixed terms like $\partial^2\phi/\partial x\partial y$. Math like physics has more different concepts than words available.

electric fields will be updated instantaneously everywhere when a charge is moved: that kind of time-dependence is *wrong*. Similarly, the Coulomb-type electrostatic potential comes from disallowing any time dependence, disallowing any but spherically-symmetric shapes, and forcing all singularities to be at an infinitesimal, extrapolated point. That adds up to quite a few arbitrary assumptions. No cookie-cutter equation can be used outside the zone of its assumptions. The most common mistakes in physics come from forgetting (or being given wrong information) about restrictive assumptions, performing exact mathematics afterwards, and generating *exactly wrong* conclusions.

3.1.1 Waves have no particular shape, unless they do

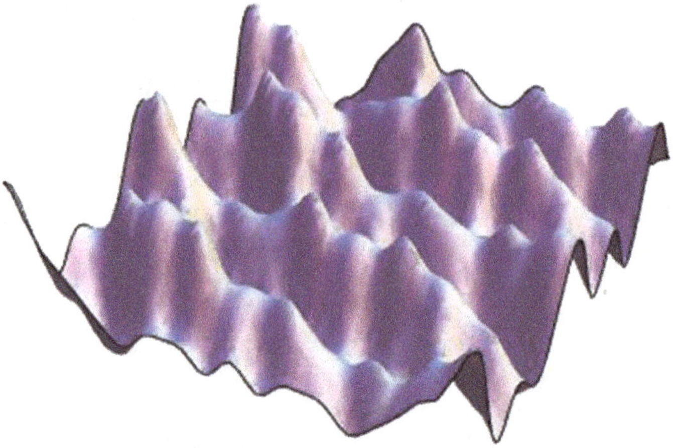

Figure 3.2. A generic wave has no particular shape. The cosine shapes are tools of human description, which work because you cannot stop people from using them, but which don't occur without a definite reason.

The fact that wave-type systems predict no particular shape is very old: d'Alembert is reported to have discovered the wave equation in 1746. It's a mistake to ask for wave equations that predict 'everything' or even formulas with *any* particular solutions. Both the wave and Poisson's equations are of the type known as *partial differential equations (PDEs)*, involving more than one partial derivatives. Such equations predict a mild but strict consistency relation upon wave shapes, and nothing more. We'll repeat this early and often.

The wave equations of quantum mechanics are PDEs, and similarly do not predict particular wave shapes. One can and will select wave shapes of particular interest for particular circumstances, yet selecting special solutions is much different from an equation predicting 'everything'. Pay attention here! The literature almost everywhere fails to say this plainly. Instead, the mathematics of PDE 'solutions' tends to express *very little information*, sometimes in *the most complicated way possible*. This is a great secret, and an amusing one. For example, the general solution to one-dimensional quantum waves in free space will be developed with

$$\psi_{\text{any}}(x, t) = \int dk\ e^{\imath kx} e^{-\imath\omega(k)t} \tilde{\psi}_{\text{any}}(k),$$
$$\text{where } \tilde{\psi}_{\text{any}}(k) = \frac{1}{2\pi} \int dx\ e^{-\imath kx} \psi_{\text{any}}(x, 0), \tag{3.3}$$

and $\psi_{\text{any}}(x, 0)$ is *any* function you chose to consider. Many find this mysterious and intimidating, and we suspect, many cruel teachers have enjoyed it. Yet the second line actually says that *any* function ψ_{any} equals ψ_{any}, and has *no information.*. The first line has information in only one (1) of all the symbols. Expressions are very complicated only because expressing *almost no information* for almost all possibilities takes a lot of notation!

Knowing that waves have no pre-arranged shape lets you bypass a few patches made to cover the mistakes of a definite shape. The term 'wave packet' was made up to correct a *wrong definition of waves in terms of cosine shapes.* To correct the wrong definition[4] a wrong principle is then invoked. The 'principle of superposition' says that you can always add two waves that are solutions and make a new wave solution. It's generally applied to make new shapes out of sums of cosine shapes, making them look more essential than ever. The 'principle' is mis-named and false in general, and because it's false, it causes problems down the road. *However*, you can add solutions to *linear* equations, which are often used in quantum mechanics.

Returning to Samuel Earnshaw's theorem of 1842: both in his time and now, there never existed an internally consistent theory of static, stable matter in equilibrium. There is no classical crystal lattice of copper atoms with happy point-like electrons bumbling amid them to make electricity 'a flow of electrons'. If a genie could assemble 10^{24} classical electrons and nuclei into an intricately chemical mole of copper, *with all electric charges equal and opposite with forces balanced out for every particle*, the smallest thermal fluctuation would cause the whole affair to *explode*.

3.1.2 Complex waves

The Schrödinger wave function is complex, with a real and 'imaginary' part:

$$\psi(\vec{x}, t) = \psi_1(\vec{x}, t) + \imath\psi_2(\vec{x}, t). \tag{3.4}$$

The notation for ψ_1, ψ_2, which are both real, is not standard. What was standard when quantum mechanics was discovered was a certain fear or superstition about complex numbers that was never justified, and by now is quite obsolete.

Complex numbers are just notation for real numbers, with *restrictions* automatically built into the addition and multiplication rules. It is a serious conceptual error to say the second part of the complex number is 'imaginary', in the sense of not existing physically. The high-school math teacher who smugly says we cannot buy $3\imath$ apples, where $\imath = \sqrt{-1}$, does not know how physics works. There are many examples in physics where pairs of real numbers match very nicely with complex

[4] Bruce McKeithan has been helpful in clarifying how the literature tends to be unhelpful.

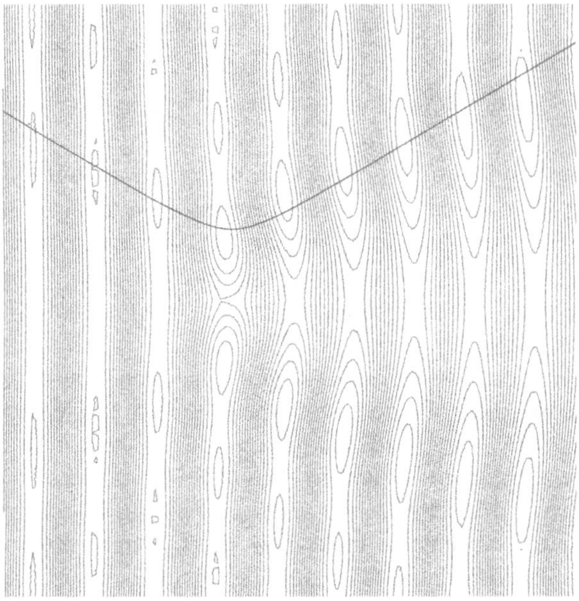

Figure 3.3. Contour map of a two-dimensional slice of the real part of the solution to the Schrödinger equation for an interaction function $V \sim 1/r$. Waves move from left to right, and roll continuously across the scattering center. A movie would be better. The curve is a hyperbola, as used in the classical Rutherford trajectory, that approximately follows the rays of wave propagation in a very rough sense.

numbers. The best known example is complex notation for resistance, inductance, and capacitance of electric circuits. The real (first) part of the wave function is the actual *wave amplitude*, although the word 'amplitude' is also used for both parts. The second part of the wave function is equally important, just as voltage and current are equally important in electric circuits. We'll explain more later: we *never* replace the complete wave with its modulus-squared $\psi^*\psi$. That is such a dreadful mistake we'll postpone even talking about it.

Complex algebra automates *rotations.* Complex $z = x + \imath y$ has exactly the same information as the coordinate pair (x, y) representing a point in a plane. Multiplying $z \to z'(t) = e^{-\imath\omega t}z$ is equivalent to the rotation

$$x \to x'(t) = \cos(\omega t)x + \sin(\omega t)y;$$
$$y \to y'(t) = -\sin(\omega t)x + \cos(\omega t)y.$$

PLEASE CHECK THIS by computing the 'real' and 'imaginary' parts of $e^{-\imath\omega t}z$. With t standing for time, the operation makes the point $(x'(t), y'(t))$ run on a circle clockwise (the direction of increasing time.) This took some planning and explains the minus sign in $\exp(-\imath\omega t)$.

Rotating a vector in the plane by 180° is equivalent to multiplying it by −1. The rotation is equivalent to the operations of two 90° rotations in sequence. In complex arithmetic $\imath$ rotates a vector by 90°, and $\imath \times \imath = -1$ rotates by 180°. No wonder that $\imath = \sqrt{-1}$! The rotations are rather transparent in $\exp(-\imath\omega t) = \cos(\omega t) - \imath\sin(\omega t)$. It apparently took a few hundred years for someone (reportedly W R Hamilton) to recognize that complex numbers were automating ordinary geometry, not imaginary non-existence. By now it is well-known, yet many *prefer to maintain* there is a mystery in complex numbers. Who benefits from it?

The algebraic properties explain the role of complex numbers in quantum mechanics. The two parts of the wave oscillate in time with phase relations much like voltage and current in AC circuits, which the complex notation automates. Consider a quantum wave with time dependence $\psi(\vec{x}, t) = \underline{e^{-\imath\omega t}\psi(\vec{x}, 0)}$. Such time dependence is exceptional, but also a fundamental building-block or thinking tool for what will be general. Separating the two parts gives

$$\begin{aligned} \psi(\vec{x}, t) &= e^{-\imath\omega t}(\psi_1(\vec{x}, 0) + \imath\psi_2(\vec{x}, 0)); \\ \psi_1(\vec{x}, t) &= \cos(\omega t)\psi_1(\vec{x}, 0) + \sin(\omega t)\psi_2(\vec{x}, 0); \\ \psi_2(\vec{x}, t) &= -\sin(\omega t)\psi_1(\vec{x}, 0) + \cos(\omega t)\psi_2(\vec{x}, 0). \end{aligned} \tag{3.5}$$

The multitude of symbols set off in the box has just the same information as the complex expression underlined . (It is also repetitive: we can't be sure you ACTUALLY CHECKED the algebra suggested.) The expression says that ψ_1 and ψ_2 rotate into each other continually, with neither being more 'real' than the other. We do not *need* complex numbers to represent quantum mechanics[5] but they are *efficient and convenient* to describe just what is wanted.

We come to the question: 'OK, so the electron wave is like two real-valued waves. Why do you need *two*?' The answer is known, and comes from the *conservation of electric charge*, as expressed by a local continuity equation (see section 4.0.4) in quantum theory. Consider a real-valued wave going like $\cos(kx)\cos(\omega t)$. That wave disappears everywhere and repeatedly at times $t_n = (n + 1/2)\pi/\omega$, for integer n. Periodically the wave reappears everywhere, out of nothing. That cannot describe a continuously conserved charge. In comparison the time-dependence of equation (3.5) shows that when one part of the electron wave goes through a minimum, the other is crossing a maximum, while there is never a moment when the wave entirely disappears. In section 6.1.4 we will see the quantum electric uses both the real and imaginary parts of the wave function.

Constructing math to match conservation laws is an everyday tool nowadays in particle physics (which is about waves of fields, not particles). For obvious reasons it is much easier to count up conserved quantities in high energy events than get down inside the little quantum waves. Conservation laws and reaction rates are enough to deduce that the 'color of a quark' needs three continuously intermixing complex waves, which tumble around like a little polarization in the hidden space called *SU* (3). It is *nothing like RGB colors* and the little joke about three colors has become an embarrassment. Understanding the matchup of math and physics of the strong interactions took about 50 years of successful and failed experiments, and the failure

[5] An enormous quasi-philosophical literature on the need for complex numbers in quantum mechanics does not recognize they are just notation for conveniently restricting operations. For example $\psi_1 + \psi_2$ never adds $Re[\psi_1] + Im[\psi_2]$. The convenient restrictions can be bypassed with quantities like $Re[\psi_1 - \imath\psi_2]$.

of all but one theories. The matchup of math and physics was *not* understood in the clumsy early days of quantum mechanics. Under the influence of the *OQT*, and to maintain its mistakes, complex quantum waves were botched in novice and high-school books, replacing the complex wave by its magnitude $\psi^*\psi$. That must *never* be done. There are *no cases* where the replacement improves understanding. The *only cases* that lead to a successful calculation are fake cases used in botched presentations.

If you want to think of one real wave, choose the 'real part' of ψ, which is certainly one valid part.

Does this mean that all quantum waves are complex? Not at all. The electromagnetic field is real-valued, because it *absolutely can* be created or destroyed. As consistent, light waves themselves have no electric charge. The absence of a conserved charge is an odd but true way of predicting that light can be emitted or absorbed. It is more fundamental than 'photons' and tells that when you get to photon waves, they will be real-valued. On the other hand, *the gluon field*, which is a beautiful generalization of the electromagnetic field, *must* carry a 'color' charge, which is conserved jointly with the quarks: that is very intricate. Naturally the gluon fields are *complex*.

Since complex functions are doubly-real, beware of a common math convention that will treat real-valued quantities as complex while 'throwing away the imaginary part'. This is never done in quantum mechanics. It is clumsy to use mathematics to generate things to be ignored or thrown away. When mathematics and physics are married well, there is nothing clumsy, and *no part* will be thrown away.

3.1.3 Wave numbers, wave vectors, plane waves

If you promise to understand that cosine-shaped waves are good math tools, *not at all* the 'solutions' nor defining facts of most waves, we'll now review the neat concept of wave numbers, symbol $\vec{k}$.

Look at figure 3.4, which shows a cosine wave with flat wave fronts and wavelength λ propagating at an angle θ to the x axis. Notice the wave is 'everywhere', covering the figure and formally extending to infinity. That is a true

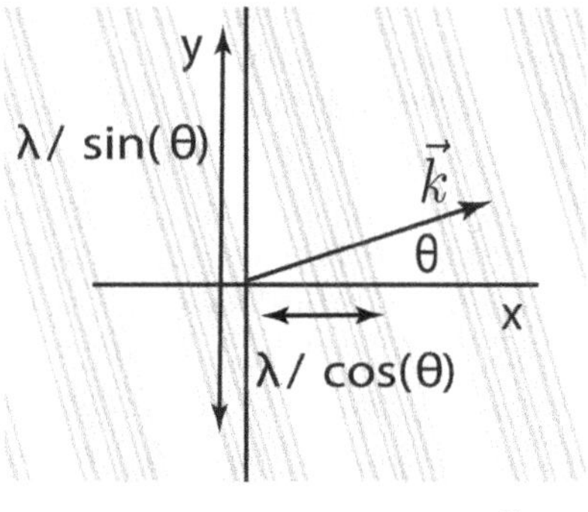

Figure 3.4. A cosine wave propagates from left to right along direction $\vec{k}$. By definition $|\vec{k}| = 2\pi/\lambda$. The *inverse* of the apparent wavelengths projected along the x axis is $k_x = |\vec{k}|\cos\theta$. This is the same with the y component k_y. Then $\vec{k} = (k_x, k_y)$ makes a true vector.

fact of cosine waves, which certainly have their limitations. If you live on the x axis, you see a periodic structure projected onto it with an apparent wavelength 'λ_x' = $\lambda/\cos\theta$. If you live on the y axis, you see a periodic structure with an apparent wavelength 'λ_y'=$\lambda/\sin\theta$. The subscript notation and a flavor of vector components prompts an idea to make a new kind of vector '(λ_x', 'λ_y')??, where ?? means there are questions. It will not work, and there cannot be a 'wavelength vector' of that kind. The x and y components of a true vector oriented at angle θ go like $\cos\theta$ and $\sin\theta$, while the 'effective wavelengths' along each axis have inverse trigonometric factors.

The solution makes a vector $\vec{k}$ out of the inverse of the apparent wavelengths. The direction of $\vec{k}$ is *perpendicular* to the wave front and in the direction the wave is moving. The vector $\vec{k}$ locally defines the direction of a *ray*. The magnitude $|\vec{k}| = 2\pi/\lambda$. The x component is $k_x = \cos\theta(2\pi/\lambda)$, and the y component is $k_y = \sin\theta(2\pi/\lambda)$. The *true vector* $\vec{k}$ is called the *wave number or the wave vector*. In three dimensions the components are $\vec{k} = (k_x, k_y, k_z)$.

The factor of 2π is included to automate 2π for convenience. Suppose $\vec{k} = \hat{x}k = \hat{x}2\pi/\lambda$. The wave is moving along the x axis. The formula for it is

$$\psi(x) = A\cos(2\pi x/\lambda) = A\cos(kx).$$

When $\vec{k}$ points in any other direction, put $\vec{k}\cdot\vec{x}$ in the argument of the cosine, and in one step you have described the wave:

$$\psi(\vec{x}) = A\cos(\vec{k}\cdot\vec{x}).$$

It's not a good idea to expand $\vec{k}\cdot\vec{x}$ or express it in trigonometric functions, because it is perfect as it stands. The *phase* of the wave is the argument of the trigonometric function. The surfaces of constant phase are the wave fronts. Those are the points where $\vec{k}\cdot\vec{x} = 0$ (say), or any constant mod-2π. Beginning at $\vec{x} = 0$, the related points $\vec{x}$ (anywhere on the plane) with $\vec{k}\cdot\vec{x} = 0$ are all points on the line perpendicular to $\vec{k}$, which are the wave fronts just as we promised.

It is not a minor point that figure 3.4 should extend everywhere, without limits. The picture frame and its arrows can be translated and put down anywhere with no significant differences. The wave is efficiently describe by the wave number $\vec{k}$, while the wave number itself has no particular position. Figure 3.5 is waiting for wave numbers on the left and right side to be drawn. Everyone discovers a momentary disorientation of not knowing where to put little arrows $\vec{k}_1$, $\vec{k}_2$. If $\vec{k}_1$ has the right length and is perpendicular to the wave fronts on the left side, it's little arrow can be put anywhere on the left. If you are given a wave number $\vec{k}_2$ for the wave on the right, it is not a concept describing the position of anything on the right. The boundary between the two regions is quite a subtle thing for wave numbers to describe: give it a try.

The visual disorientation of looking at a plane wave comes from very sophisticated human visual processing that reads more into the picture or the math expression than the actual information that is present. The last thing to suspect would be any importance to the unlikely, silly extension of the plane wave to actually exist *everywhere*. One needs a sense of scale. The typical wavelength of

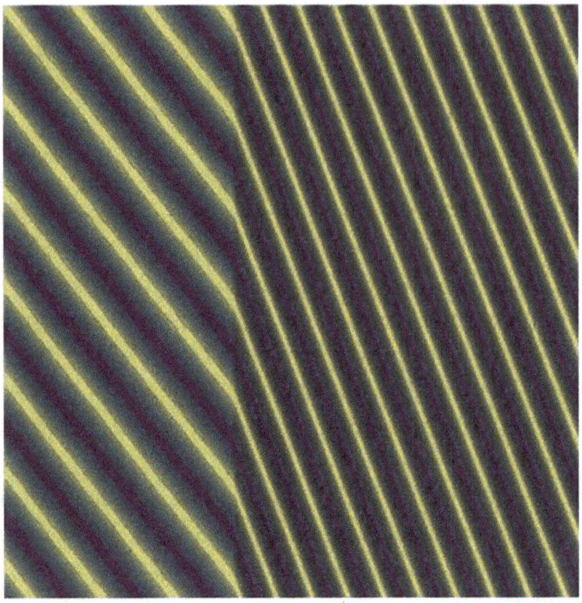

Figure 3.5. Refraction of a plane wave. The directions of two wave numbers $\vec{k}_1$, $\vec{k}_2$ are clear in each zone. Yet precisely where to locate a wave number is quite ambiguous, because wave numbers have no particular position.

protons in the beam of the Large Hadron Collider is about 10^{-17} cm. That's a hundred million waves per nanometer. For the purposes of engineering accelerator physics, a cubic nanometer is infinitely small, and for the purpose of exploring quarks it is infinitely large. While plane waves never exist they are the constant tool of simplified description. When more localized wave configurations are described using plane waves, the *uncertainty relation* provides a rough estimate of the spread in wave numbers for a given spread in spatial size. It is a math fact about superpositions of waves that has no physical information.

Opening a quantum mechanics book, or almost *any* physics book on a subject newer than thermodynamics, will find hordes of formulas with $\exp(\imath\vec{k}\cdot\vec{x})$. When someone says 'this is a plane wave' it sounds like 'a plain wave' while it looks weird and not plain. Those not instructed in the art of the wave vector will have two conflicts.

First, the imaginary conflict. Since $\exp(\imath\vec{k}\cdot\vec{x}) = \cos(\vec{k}\cdot\vec{x})) + \imath\sin(\vec{k}\cdot\vec{x})$, one complex exponential makes available cos and sin configurations. Multiplying by $\exp(\imath\delta)$ induces a phase shift $\vec{k}\cdot\vec{x} \to \vec{k}\cdot\vec{x} + \delta$. (Try it.) Such things as a complex index of refraction fall into place with lovely consistency of math and physics. The unknown genius who made so much so easy should have been remembered[6].

Moreover, the exponential automates many trigonometric formulas, such as

$$e^{\imath\vec{k}_1\cdot\vec{x}} \times e^{\imath\vec{k}_2\cdot\vec{x}} = e^{\imath(\vec{k}_1+\vec{k}_2)\cdot\vec{x}}. \tag{3.6}$$

[6] James MacCullagh, FRS 1809–47.

When there is a *product* like this the wave numbers *add*. There are real surprises in the consequences of THIS easy formula! As usual, complex numbers are not hoodoo[7] but a friend automating operations of two real numbers.

Next, if you lack skill or warning, you might forget that the phase or argument of the trigonometric functions is $\vec{k} \cdot \vec{x}$, so that $\vec{k} \cdot \vec{x} = 0$ finds all the points which are constant-phase wave fronts passing through the origin, (to repeat!) and all then perpendicular to $\vec{k}$. That surface in three dimensions is a *plane*, for goodness sake. If you translate to another point $\vec{x}_*$, then $\vec{k} \cdot (\vec{x} - \vec{x}_*)$ will again give all the points of a plane perpendicular to $\vec{k}$ and going through $\vec{x}_*$. After you know this, you must transfer your attention to $\vec{k}$, which is so easy, while remembering the *plane* of constant phases it controls is absolutely *huge*.

If you lack skill or warning, or you are brilliant, you might decide to expand the terms in the exponent of the plane wave. Unless you have a reason, *never expand a dot product*, nor replace it by a formula involving magnitudes and relative angles. You will make things longer, worse, and more difficult to interpret. But if you disobey the plane wave, you discover

According to the wave theory of light, the light rays, strictly speaking, have only fictitious significance. They are not the physical paths of some particles of light, but are a mathematical device, the so-called orthogonal trajectories of wave surfaces, imaginary guide lines as it were, which point in the direction normal to the wave surface in which the latter advances (cf. Fig. 3 which shows the simplest case of concentric spherical wave surfaces and accordingly rectilinear rays, whereas Fig. 4 illustrates the case of curved

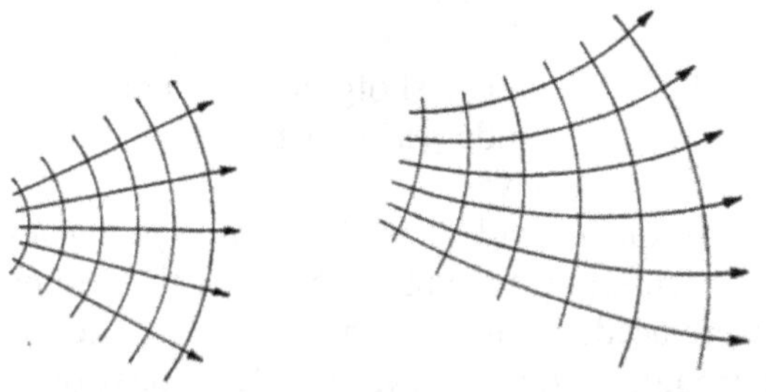

Fig. 3. Fig. 4.

Figure 3.6. Curved wave fronts illustrated by Schrödinger in his 1933 Nobel Prize lecture. His reference to 'imaginary guide lines' describing *rays* of light with only 'fictitious significance' refers to the notion of particle trajectories. © The Nobel Foundation 1933.

[7] There is a transitive verb, where one can be hoodooed by math teachers, so watch out.

$$e^{i\vec{k}\cdot\vec{x}} = e^{ik_x x}e^{ik_y y}e^{ik_z z}. \tag{3.7}$$

The sum in the exponent becomes a product. Equation (3.7) is one of the exceptions where disobedience becomes a power tool. Many calculus-driven operations such as integrals and integral transforms work smoothly on products of functions. Since that's not our topic we'll move on, with the information that *what seems to be a hundred math tricks is actually ten thousand or more math tricks*, almost all of which is only a *half-dozen* or so very good tricks in different combinations, which will accumulate with experience: and you have just seen one of them.

3.1.4 Photons and other waves are never localized at points

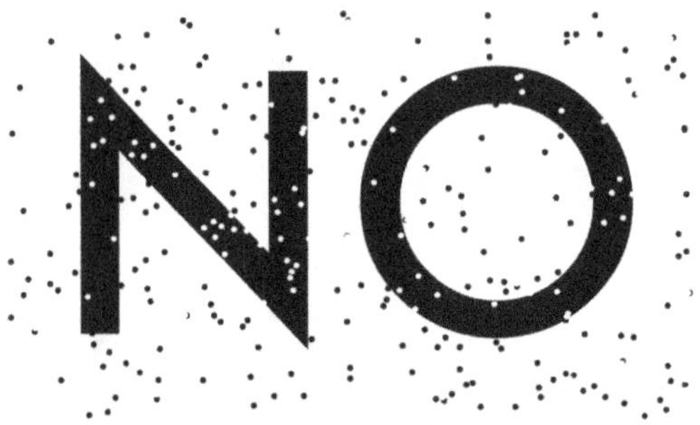

No matter what we write, someone will expect us to say that light and electrons were later found to be both a wave and a particle, and neither, all at the same time, which cannot really be understood. To repeat our message, we painted it on a wooden board, set up against an NRA-disapproved backstop in a western US state, and decorated it with a double-barreled shotgun[8]: the particle-like photon was the center-point of a 20-year propaganda period between 1905 and 1925. It was the idea of the 'photon' that did not survive.

By 1935 high-quality theory had *derived from first principles* all the phenomena the 'old' photon had been set up to explain. None of the ad-hoc assumptions of the old photon theory were needed. The new theory showed they were not correct. The old ideas died. We are not writing to resell the dead ideas of dead people.

Quantum mechanics was first developed for electrons, and then applied to electromagnetic fields. The combination is called quantum electrodynamics (QED). A signal that QED was on track came when the photon was *derived* as a particular output of the theory. If there had never been an old quantum theory, the photon would be a prediction, and also a prediction without the mistakes of the *OQT*. The internal structure of the photon, which is a lovely creature, turns out to be a 'quantized ray'. It is an entirely delocalized resonance of the electromagnetic field.

[8] Conducted at 30 yards with a single Winchester 20-gauge #9 dove and quail load, modified choke, from horseback.

They are never short, they are as long as the system making them can make them long. Still there's a clear and definite importance of a whole number of photons.

3.2 Fundamental constants without the kilogram

Many basic physical relations can be anticipated by dimensional analysis. In 1914 Edgar Buckingham [5] perfected it as a *predictive* approach to physics, as opposed to the more basic idea that dimensions should be consistent. This followed 50 years of observations by Fourier, Maxwell, Rayleigh, and others that the system of units matters, that it is *arbitrary*, and that consistency can be very restrictive.

Buckingham's example. Buckingham wrote: 'We desire the thrust of a ship's propellor as a function of the propellor's shape, size, the density of water, and the ship's speed…', given certain dimensionless shape ratios. This seems to be a difficult problem. The dimensionful quantities are the propellor radius r=*length*, the Newtonian mass density $\rho = mass_N/length^3$, the speed $v = length/time$ and force $f = mass_N \times length/time^2$. Only one combination has the dimensions of *force*:

$$f = \Pi_B\, r^2 \rho^1 v^2. \tag{3.8}$$

Here Π_B is a dimensionless constant, which Buckingham's method isolates, and which must absorb the shape factors of the propellor, the details of water and hydrodynamics, and so on. Whatever that constant is, the method has made a prediction, as if by magic, of the more important *relationships*. Since the objective was the Newtonian concept of *force*, the inputs needed a quantity like the Newtonian $mass_N$ density that referred to Newtonian theory. Soon we'll discover that Newtonian mass is a 'dirty word'.

Example: the blue sky. Rayleigh discovered the sky is blue because higher frequencies scatter off atoms in proportion to the illuminating frequency ω^4. The atmosphere is about 100 km deep, which is just right for backscattering to happen most of the time for blue light with angular frequency near $\omega_{\rm blue} \sim 4 \times 10^{15}\ {\rm s}^{-1}$. This is rather finely tuned, because the scattering is 10^4 times smaller for 10 times smaller frequency: red reflects back very little from the sky. By dimensional analysis, the Rayleigh scattering cross section (an effective area) must then go like $\sigma_{Rayleigh} \sim d^6\omega^4/c^4$, where d is a typical diameter of the scattering atom. The *interaction* is dynamical and the effective area is not $\pi d^2/4$. (And this is why '*parcle* cross sections' do not measure *parcle* size, hey!) Also notice that *no concept of Newtonian mass or force is involved.* From the color of the sky Rayleigh was able to deduce that atoms are close to 10^{-8} cm in size, although the d^6 dependence left some uncertainty. Reversing the argument with information that molecules are close to 10^{-8} cm in size predicts the sky is blue[9] on Earth, and white to yellow (the color of the Sun) on other habitable planets with sufficiently deep atmospheres that all the

[9] It is blue during the day. We don't know what color it is at night.

colors bounce back. Since Mars has a very thin atmosphere, the stars probably can be seen in the daytime, and the atmosphere will be[10] *deep purple.*

Continuing. In Buckingham's time it was assumed that Newtonian physics and the Newtonian *mass* unit was necessary to do physics. *After* quantum mechanics was discovered Newtonian physics and its kilogram became irrelevant for fundamental physics. It will be be maintained for engineering, commerce, and classical approximations, but there is no role for it at the fundamental level.

The necessity of the *kilogram* in Newtonian physics comes from a gap in the equation $\vec{f} = m\vec{a}$. While the acceleration $\vec{a}$ is a 'space–time' observable, the equation introduces two quantities $\vec{f}$ and m to measure one thing. Measuring one component $a_x = f_x/m$ cannot determine f_x or m separately. Newtonian theory has a lapse of incomplete definition, which was covered by distributing arbitrary force meters or copies of the kilogram to patch the gap. Everything downstream of $\vec{f} = m\vec{a}$ inherits the weakness: in particular, Newtonian *energy* has a faulty definition (see section 8.2.1), and faulty units inherited from the *kilogram.*

3.2.1 Getting rid of the kilogram

Everyone agrees the kilogram is a human convention. Could quantum theory be done without a Newtonian mass unit entirely? To explore this, imagine we began with quantum mechanics and no prior notion of the Newtonian mass unit. We would need to do physics and find our fundamental constants from data without using the *K* part of *MKS* units. To make this more interesting, let's temporarily restrict attention to experimental data known before 1901, and avoid all bias of the *OQT*. We do not restrict theory concepts to coming before 1901: we imagine quantum mechanics had been understood *without* the digression of 'quantum *parcles*'.

We will have units of time and length. The unit of time suffices to define the unit of length when we adopt the *theory* that the speed of light c is a universal constant. That was done in the late 20th Century by fixing $c = 2.99\ldots\times10^8$ m s^{-1} to a reference value. That required a highly precise definition of the *second*, which was adopted in terms of very technical atomic physics (rubidium). For conceptual purposes the 19th Century could have defined the second in terms of the lowest vibrational frequency of the hydrogen atom, $\omega_H = 2.08 \times 10^{16}$ s^{-1}, a number known from Rydberg's work in the 1880s.

And so the 19th Century already knew that the equations of quantum mechanics *must* have a constant with dimensions of frequency, to predict an atomic frequency. (It was specifically noticed by James Jeans in 1904; see section 6.1.2.) The electron wave equation indeed has a fundamental frequency constant $\omega_e \sim 7.8 \times 10^{20}$ s^{-1}. An alternative, human scale parameter is $\mu_e = \omega_e/c^2 = 0.87$ s cm^{-2}. This fact may be new to you, but remember, we are imagining a different history without the kilogram.

The ratio of ω_H/ω_e is dimensionless, and must come from the dynamics of the atom. The quantum theory developed in section 7.2.1 shows that $\omega_H/\omega_e = \alpha^2/2$, where α is the dimensionless coupling of the electromagnetic interaction.

[10] Some photos from Mars rovers show a reddish sky, but that's dust. Other photos do look purple.

Table 3.1. Fundamental constants α, ω_e, ω_p and experimental data predicted by them. The fundamental eletromagnetic coupling constant α is called the 'fine structure constant' for historical reasons. Number densities n refer to typical solids. The column *oom estimate* shows numbers estimated to order of magnitude. More precise determination of fundamental constants involve global fits to high precision data sets.

name	*symbol*	*formula*	*oom estimate*
fine structure constant	α	$2a_0\omega_H$	$0.0073 \sim \frac{1}{137}$ (fitted)
electron frequency	ω_e	$1/2a_0^2\omega_H$	$7.8 \times 10^{20}\ \mathrm{s}^{-1}$(fitted)
proton frequency	ω_P	$1835\omega_e$	$1.4 \times 10^{24}\ \mathrm{s}^{-1}$ (fitted)
hydrogen frequency	ω_H	$\alpha^2\omega_e/2$	$2.07 \times 10^{15}\ \mathrm{s}^{-1}$
hydrogen size	a_0	$c/(\alpha\omega_H)$	0.529×10^{-9} cm
atomic density	n_A	$a_0^{-3} \sim An_P$	$10^{24}\ \mathrm{cm}^{-3}$
electron density	n_e	$\sim n_A$	" "
Fermi frequency	ω_F	$n_e^{2/3}c^2/\omega_e$	$1.2 \times 10^{16}\ \mathrm{s}^{-1}$
Fermi pressure	$\omega_F n_e$	$n_e^{5/3}c^2/\omega_e$	$1.2 \times 10^{40}\ \mathrm{s}^{-1}$
plasma frequency	ω_{plasma}	$\alpha\sqrt{n_e c^3/\omega_e}$	$1.6 \times 10^{16}\ \mathrm{s}^{-1}$
sound velocity, solids	v_{Ss}	$\sqrt{\omega_F n_e/n_A\omega_A}\, c$	$1.1 \times 10^8 \frac{1}{\sqrt{A}}\ \mathrm{cm\ s}^{-1}$
sound velocity, gas	v_{Sg}	$\sqrt{\gamma T_\omega/\omega_{Air}}\, c$	$331\sqrt{(T_\omega/3.6 \times 10^{13}\ \mathrm{s}^{-1})(28/A)}\ \mathrm{m\ s}^{-1}$
Rayleigh cross section	$\sigma_{Raleigh}$	$\alpha^2\omega^4 a_0^6/c^4$	$1.5 \times 10^{-32}\ \mathrm{cm}^2(\omega/10^{16}\ \mathrm{s}^{-1})^4$

(Everyone uses this standard approximation: a correction of order 10^{-3} will be discussed soon.) The characteristic size of the hydrogen atom called $a_0 \sim 5.29 \times 10^{-9}$ cm, which is very precisely defined by the theory, is $a_0 = \alpha c/\omega_e$, where c is the speed of light, which was already known to 1% accuracy by the 1860s. (Check the numbers. If you know the atomic size and α, you can predict ω_e.) The first clue to atomic size is the typical number density of atoms in solids, $n_A \sim a_0^{-3} \sim 10^{24}\ \mathrm{cm}^{-3}$, as discussed further in section 2.3. From experiments, the atomic size and atomic frequency determine two fundamental constants $\alpha \sim 0.0073$ and $\omega_e = 7.8 \times 10^{20}\ \mathrm{s}^{-1}$.

While not done until later in history, the two constants α and ω_e (plus one more) suffice to predict everything in table 3.1. Many of the quantities have a Newtonian formula. Since there is no classical theory of matter, the Newtonian formulas have an ad-hoc 'fudge factor' fit to experiment case by case. For example, the Newtonian formula for the velocity of sound is $v_S = \sqrt{K/n_A m_{NA}}$, where K is called the bulk modulus of the material, n_A is the number density of atoms, and m_{NA} is the Newtonian mass per atom. The fudge factor K is defined circularly from Newtonian mass and the speed of sound. This is somewhat subtle, due to the formula hiding how the formula was obtained. An engineer might say K is found by force or pressure meters. The fudge factor K comes from outside the Newtonian theory, no matter how you get it.

3.2.2 Quantum theory is ambitious

The quantum theory is much different, and seeks to predict everything. The fundamental constants of α and the fundamental frequency constant of electrons (ω_e), protons (ω_P), or entire atoms (ω_A) are the only parameters allowed. Since we no longer have a use for Newtonian mass, the frequency parameters must provide the dimensional factors, while providing predictive power from the theory.

The proton's constant ω_P can be estimated from the speed of sound v_{Ss} in typical solids, as follows. The *Fermi pressure* of degenerate electrons predicts most of the classical bulk modulus. The Fermi pressure is the Fermi frequency $\omega_F = n_e^{2/3}c^2/\omega_e$ per volume, where n_e is the electron number density. This is an absolute quantum prediction, often using a free-electron approximation, but in principle (and in the 21st Century) computable for arbitrarily precise electronic and crystal structure information. It is *not* a fudge factor. Then $v_{Ss} = \sqrt{\omega_F n_e c^2/\omega_A n_A}$ is dominated by dimensional analysis, where $\omega_A = A\omega_P$ is the frequency parameter of atoms with atomic number A propagating the sound wave. There are typically a few valence electrons per atom so $n_e/n_A \sim 1$. Then $v_{Ss} \sim \sqrt{n_e^{2/3}c^2/A\omega_e\omega_P}$. By algebra

$$v_{Ss}^2 \sim \left(\frac{10^{16}\text{cm}^{-2}}{7.8^2 \times 10^{40}\text{s}^{-2}A}\right)\left(\frac{9 \times 10^{20}\text{cm}^2}{\text{s}^2}\right)\left(\frac{\omega_e}{\omega_P}\right)c^2;$$

$$\frac{\omega_P}{\omega_e} \sim 900\left(\frac{56}{A}\right)\left(\frac{5000\text{m s}^{-1}}{v}\right)^2.$$

For iron $A = 56$ and $v_{Ss} = 5000$ m s^{-1} gives $\omega_P \sim 900\omega_e$, which 'weighs the proton' on the frequency scale. For diamond $A = 12$ and $v_{Ss} = 12\,000$ m s^{-1}, which predicts $\omega_P/\omega_S = 900(56/12)(5000/12\,000)^2 = 1750$. Given the effects neglected, the order of magnitude is $\omega_P \sim 10^3\omega_e$.

A much better determination is done with precision atomic spectroscopy. Since the electron and proton interact, the spectral frequencies of the physical hydrogen atom depends on a combination of electron and proton frequencies (equation (10.3), section 10.1.1):

$$\omega_H = \alpha^2 \frac{\omega_e\omega_P}{2(\omega_e + \omega_P)} \sim \frac{\alpha^2\omega_e}{2}\left(1 - \frac{\omega_e}{\omega_P} + \cdots\right).$$

This reduces to the commonly remembered relation $\Omega_H = \alpha^2\omega_e/2$ when $\omega_e \ll \Omega_P$. A sufficiently precise measurement of ω_H and another precise measurement of ω_e or ω_P predict the ratio. Using modern numbers gives $\omega_e/\omega_P \sim 0.00054$, or $\omega_P/\omega_e \sim 1835$.

3.2.3 Mass ratios

The number above is the same ratio as the Newtonian proton to electron mass ratio:

$$\frac{\omega_P}{\omega_e} \sim \frac{m_{NP}}{m_{Ne}}. \tag{3.9}$$

This is expected from a rough argument called Ehrenfest's theorem; see section 8.1.4. It is remarkably crude: the quantum waves are replaced by averaged quantities that are plausibly related to the imaginary ray trajectories shown in Schrödinger's figure 3.6. It is an uncontrolled approximation that is clearly *not* exact.

The reliability of the relation above is not well understood. The precision of Newtonian physics in quantum circumstances never cleared up, except that it failed. This is not controversial: nobody knows what would be meant by 'the Newtonian mass of a quantum electron'. With few exceptions, 21st Century quantum mechanical parameters are measured known to high precision using quantum mechanical equations and observables. The left hand side of equation (3.9), which *is* very precisely measured, can be used to *define and predict* the right hand side[11], for any Newtonian quantities that might be wanted. At least that is how history *should have* gone.

Unfortunately, history went the other way. There was a blind conviction from 1900–25 that Newtonian mechanics was 'established' and its definitions of fundamental constants were 'exact and final'. When Bohr, Born, Heisenberg, and so on started questioning Newtonian physics, we see no record of them questioning the fundamental constants of Newtonian physics. That is surely due to a misconception that fundamental constants are self-defining, and determined[12] by 'experiments'. Actually fundamental constants get their meaning from a *theory*, and cannot be measured without a *theory* to define them.

The traditional presentation of quantum mechanics used the *old theory* as defining concepts. The 'Newtonian mass of a quantum electron' was accepted without discussion. To this day one finds surprisingly careless use of the 'mass of the electron' in terms of a Newtonian $mass_N$ reference parameter $m_{Ne} = 9.1 \times 10^{-31}$ kg, which puts *kilogram* units back into a quantum theory that does not need them. When the most basic system of units and measurements (wrongly) pre-supposed Newtonian definitions, it worked like a chronic plague to maintain the mistakes *after* the new quantum theory was discovered.

Old units and new units. There are two ways to proceed. The pre-quantum tradition assumed Newtonian mass was fundamental, and then used Planck's constant to cancel out *MKS* units where they don't appear in quantum theory, which happens to be everywhere. To use the *MKS* system replace frequencies $\omega_e \to m_{Ne}c^2/\hbar$, $\omega_P \to m_{Pe}c^2/\hbar$, and so on in kilogram units. This maintains $\omega_e/\omega_P \to m_{NP}/m_{Ne}$, and all other ratios. The numbers of the *MKS* system have an arbitrary scale, because the *kilogram* is an arbitrary unit[13] invented in 19th Century France. Since ω_e has no reference to a kilogram, then m_{Ne} in kg and $\hbar$ in units of kg-m^2 s^{-1} cancel out the *kilogram*. The unacceptably large errors of Newtonian mass and $\hbar$ (defined in terms of it) wreak havoc in high precision work, but the replacements are OK for people wanting *MKS* numbers.

---

[11] Defining the Newtonian mass in terms of observable frequencies closes the logical gap of a 300 year old theory. But since we're using quantum mechanics, we expect Newtonians to remain uncertain.

[12] A surprising number of claims about 'proven facts of experiments' are actually *bluffs*. Bluffs can be detected by asking for the experimental uncertainties. It is rare to find a $bluff \pm \Delta bluff$.

[13] It distresses some to realize that Planck's constant is not found in nature but instead found in the *kilogram*. The information is pretty useful for destroying the *OQT*.

The other way to proceed uses ω_e in frequency units, or an equivalent 'quantum mass parameter' $\mu_e = \omega_e/c^2$. This kind of mass is a frequency called a mass for everyone wanting to keep the word 'mass'. Then calculations of spectral frequencies and time evolution are predicted without any mention of the *kilogram*, just as the table shows.

Example: light predicts sound. Planck found[14] that the frequency spectrum of light goes like $\nu^2/(e^{\nu/T_\nu} - 1)$, where T_ν is the temperature in units of frequency. The peak of the distribution is at $1.59\ T_\nu$. This expression bypasses the Newtonian temperature in degrees Kelvin, which is redundant. For example, a black body at temperature 1000 K is 'red hot', emitting considerable radiation near $10^{14\pm1}\ \mathrm{s}^{-1}$, and looking at the color tells you it is 'hot' without a thermometer. The arbitrary unit of the Kelvin is defined by a conversion factor $1°\,K = 1.31 \times 10^{11}\ \mathrm{s}^{-1}$. Then 0 °C = 273 K = $273 \times 1.3092 \times 10^{11}\ \mathrm{s}^{-1} = 3.57 \times 10^{13}\ \mathrm{s}^{-1}$. Looking at ice in the infrared regime will 'see the temperature of ice', *entirely bypassing the kilogram*.

The ideal gas law is the same law in T_ω units. The speed of sound in an ideal gas is $v_{Sg} = \sqrt{\gamma T_\omega/\omega_{Air}}\, c$, where ω_{Air} is the molecular frequency (mass) parameter of air, and $\gamma = 7/5$ accounts for isentropic compression. The sound speed is 331 m s^{-1} for air at 273 K. Solving for ω_{Air} gives

$$\omega_{Air} = \frac{1.4 \times 3.57 \times 10^{13}\ \mathrm{s}^{-1}}{331^2\ \mathrm{m}^2\ \mathrm{s}^{-2}} \frac{9 \times 10^{16}\ \mathrm{m}^2}{\mathrm{s}^2} = 4.11 \times 10^{25}\ \mathrm{s}^{-1},$$

$$\frac{\omega_{Air}}{\omega_P} = \frac{4.11 \times 10^{25}\ \mathrm{s}^{-1}}{1.43 \times 10^{24}\ \mathrm{s}^{-1}} = 28.7.$$

Air is close to 80% N_2 (molecular weight 28) and 20% O_2 (molecular weight 32), averaging to molecular weight 28.8, which agrees. The speed of sound directly measures the mass of air molecules in units of s^{-1}. Alternatively, $\mu_{Air} = \omega_{Air}/c^2 = 45\,500\ \mathrm{s\ cm}^{-2}$.

Example: the gas constant. The practice of making and distributing physical copies of the kilogram is archaic and imprecise. Scientists have long sought an alternative based on independently reproducible measurements. The kilogram unit enters the ideal gas law $PV=RT$ through the molar gas constant $R = 8.314\,45\ \mathrm{J\ mol}^{-1}\mathrm{K}^{-1}$ One mole of gas ($N_A = 6.02 \times 10^{23}$ molecules) contained in 22.4 liters of volume at standard temperature 273.15 K has an absolute pressure $P = 101.325$ Pa. Applying that pressure to a cylinder of area $A = 0.967\,84\ \mathrm{cm}^2$ makes a force $f = 9.8066$ N. The standard acceleration of gravity happens to be $9.806\,65\ \mathrm{m\ s}^{-2}$. The pressure will balance a piston of mass 1 kg compressing the gas under the force of gravity, according to the force balance equation $mg = PA = RTA/V$. That will serve to experimentally define a standard kilogram. The temperature in frequency units is $3.58 \times 10^{13}\ \mathrm{s}^{-1}$, or equivalently the molar gas constant $R = 2.155 \times 10^{37}\ \mathrm{K}^{-1}$. The kilogram is then determined by

[14] What Planck found is literally correct. What he *reported* was converted to *MKS* units using the Boltzmann constant (another of Planck's inventions) from other data to define 'temperature' in *MKS* units. The low-frequency limit found by Rayleigh and Jeans also suffices for the relation.

The gas constant R in a number of unit systems listed in *Wikipedia*, along with natural units. Uncertainties are the values of last digits in parentheses. Actually seven more unit conversions were provided. Silly, right?

Value of R	units
8.314 459 8(48)	kg m^2 s^{-2} K^{-1} mol^{-1}
8.314 459 8(48)	J K^{-1} mol^{-1}
8.314 459 8(48) × 10^7	erg K^{-1} mol^{-1}
8.314 459 8(48) × 10^{-3}	amu (km^{-1} s)$^{-2}$ K^{-1}
8.314 459 8(48)	m^3 Pa K^{-1} mol^{-1}
8.314 459 8(48) 10^6	cm^3 Pa K^{-1} mol^{-1}
8.314 459 8(48)	L kPa K^{-1} mol^{-1}
8.314 459 8(48) × 10^3	cm^3 kPa K^{-1} mol^{-1}
8.314 459 8(48) × 10^6	m^3 MPa K^{-1} mol^{-1}
8.314 459 8(48)	cm^3 MPa K^{-1} mol^{-1}
8.314 459 8(48) × 10^{-5}	m^3 bar K^{-1} mol^{-1}
8.314 459 8(48) × 10^{-2}	L bar K^{-1} mol^{-1}
2.155 × 10^{37}	mol^{-1} K^{-1}

$$1\ \text{kg} = N_A TA/Vg,$$

$$= \frac{6.02 \times 10^{23} \times 3.58 \times 10^{13}\ \text{s}^{-1} \times 0.967\,84\ \text{cm}^2}{22.4 \times 10^3\ \text{cm}^3 \times 980.665\ \text{cm s}^{-2}} = 9.495 \times 10^{29}\ \text{s cm}^{-2}.$$

Recall the electron frequency parameter, which comes directly from quantum mechanical data, is $\mu_e = 0.87$ s cm^{-2}. Converting units gives

$$\mu_e = \frac{0.87\ \text{s cm}^{-2}}{9.5 \times 10^{29}\ \text{s(kg}^{-1}\,\text{cm}^{-2})} = 9.1 \times 10^{-31}\ \text{kg}.$$

Comment. It is interesting that the magnificent technology of the 21st Century coexists with the anachronisms of the *kilogram* and the practice of reporting meaningless conversion constants on the same footing as true fundamental constants. Table 3.2 from *Wikipedia* shows a number of high-precision values for the gas constant R. The table suggests that Wikipedia readers lack confidence in converting units. (Let's hope we typed correctly!) The value of R is an example of those 'experimentally defined' fundamental constants not only based on outmoded theory, but *insisting on definitions* from outmoded theory. The gas constant R describes the properties of ideal gases. There exist no ideal gases. To 'experimentally measure' the gas constant the speed of sound is measured in argon as a function of pressure and extrapolated to zero pressure. The eight-digit numbers represent very expensive high-precision measurements of a gas that does not *exist*.

3.2.4 The identity of energy and frequency

When quantum mechanics was discovered most physicists were experts in *generalized* classical mechanics in Hamiltonian and Lagrangian form. Scholarship went downhill since. As a result, students trying to learn quantum theory are set to be waylaid by concepts from classical mechanics (Poisson brackets, generators of canonical transformations, Hamiltonians, etc) they are *not expected to understand.* As Feynman must have said, you can fool your friends, and fool your teachers, but you cannot fool physics, which will clobber anyone not understanding it. (Which is another reason why insisting quantum mechanics *would not be understood* once made things so difficult.)

We've structured our presentation to avoid that history. The Schrödinger equation will calculate frequencies measured, and that's all you really need. Still the concepts and facts of Hamiltonians are important, and very strongly recommended after your first understanding of quantum mechanics has been developed. Once you study it all deeply, there is a concise summary:

The frequency operator is the Hamiltonian operator is the frequency operator, and it will be whatever you choose for your model.

Henceforth, and when translating sources, you can use 'Hamiltonian' and 'frequency operator' interchangeably.

Also here in small print is information in case you investigate the digression:

Many unfortunate textbooks[15] will tell you that Hamiltonian and Lagrangian mechanics are mere re-formulations of Newtonian mechanics, because the book authors don't know any better (*badkab*). After botching classical Hamiltonians, the road will be golden to botching Hamiltonian operators.

When you study it, a few useful facts will convincingly show that Hamiltonian physics is different and profound. *Fact* 1: the Newtonian definition of 'energy' comes from work, and depends on a path C via a line integral $\int d\vec{x} \cdot \vec{f}$. Right? Hamiltonian energy *never* depends on a path, and is always given as a definite *function. Fact* 2: the Newtonian definition of 'momentum' is $\vec{p}_N = m_N\vec{v}$. The Hamiltonian use of momentum p_i is not a definition, it is a set of independent *coordinates* for a system, effectively doubling the phase space of generalized coordinates called q_i. The formula relating p_i to other variables is not determined until the system is defined. There are *no cases* where the two definitions of p_i and $\vec{p}_N$ coincide[16] *except* for a circular-defined Hamiltonian model rigged to reproduce Newtonian physics as a special case. *Fact* 2.5: Hamilton's equations predict time evolution:

$$\dot{q}_i = \partial H/\partial p_i; \qquad \dot{p}_i = \partial H/\partial q_i. \tag{3.10}$$

Choose any function $H(q_i\, p_i)$ you desire for your 'Universe'. Hamilton's equation will reveal what Universe you just made. *Fact* 3: guesswork and fitting data has produced many excellent Hamiltonian models, defined by particular functions $H(q_i, p_i)$. They explain experimental data. Meanwhile, people have been looking

[15] *Thornton and Marion* is an example.[6]

[16] The best selling, most difficult, and most devastatingly harmful textbook by Goldstein [7] writes on the first page that the definition of momentum $\vec{p} = m\vec{v}$, contradicting the generalized Lagrangian and Hamiltonian mechanics the book is about. Many prefer Landau's book, which is rather cruel, but at least much *shorter*.

for a perfectly Newtonian system for 300 years, and never found one that worked in all detail. *Fact* 4: the books of the *badkab* will tell you the *definition* of the Hamiltonian is $H = KE + PE$, where PE is the potential energy, and $KE = \vec{p}^2/2m_N$ is the Newtonian kinetic energy $m_N\vec{v}^2/2$ rewritten using $\vec{p}_N$. This is a bad definition and false in general. There are *no cases* where it is true, *except* for a circular-defined Hamiltonian model rigged up to reproduce Newtonian physics as a special case. Yet every day millions of students are memorizing a wrong 'fact of the Universe' that energy must always be the sum of $KE + PE$. We hope these *facts* inspire you to study.

Schrödinger learned Hamiltonian mechanics from his dissertation advisor, Friedrich Hasenöhrl (1874–1914) who Schrödinger claimed would have discovered quantum mechanics if he had not died early. Schrödinger knew mechanics backwards and forwards, including the 1840s discovery by W R Hamilton (1805–65) that classical particle trajectories could be computed by a method of wave optics and 'rays'. If you study Einstein's work, you will see he used Hamiltonian and Lagrangian physics extensively. Einstein also tended to repackage arguments with meter sticks, clocks, and moving railroad cars that didn't give it away.

After Schrödinger discovered his equation he commented in a letter to Einstein about *the identity of frequency and energy in quantum theory*. The link comes from the Lagrangian–Hamiltonian quantity called *action*. Bypassing a one semester course we strongly recommend... the action $S(\vec{x}, t)$ has all the information about a physical system *including* its initial conditions. The variables that extremize the action solve the equations of motion. That is the content of the 'principle of stationary (or least) action', $\delta S = 0$, which is super-sophisticated notation if nothing else[17]. Then *given* the function $S(\vec{x}, t)$ representing solutions the Hamiltonian is defined $H = -\partial S/\partial t$. The *energy* is the numerical value of the Hamiltonian. These definitions of Hamiltonian and energy are general, and universal. They do not depend on the details of the system. They are 100 years more advanced than Newtonian physics.

When Planck had first connected frequency to energy, Einstein knew the energy referred to $H = -\partial S/\partial t$ and the frequency was about the action. But the action of what? *After* Schrödinger found his wave function, there was an answer. The formula $H = -\partial S/\partial t$ reduces to numerical values $E\psi \to \imath\partial\psi/\partial t \to \omega\psi$ when $\psi \sim e^{-\imath\omega t}$. Einstein's relation was true for frequency eigenstates, and also circular for that case: it is not true for more complicated waves. Einstein was off track for 25 years thinking about *parcles*. The action of the Schrödinger equation is not very exotic, and has a term $S = \imath\int d^3x\, \psi^*\partial\psi/\partial t$. Put in $e^{-\imath\omega t}$ and $\partial S/\partial t \sim \omega$. (More information is in section 8.2.1.) The action formula is general and predicts *exactly when* energy equals frequency: proposing energy IS frequency was not general and could never predict the action. *After* Schrödinger published his equation, Einstein

[17] Since $\delta S = 0$ defines the equations, action is dimensionless, and the attempt to give it units just cancels out with '0' on the right hand side. The 'cult of the quantum of action' $\hbar$ was an idea that would fail, which failed.

heard a garbled report from Planck, where apparently the equation was reported wrong. Einstein then sent Schrödinger a suggested correction[18] for what a viable wave equation ought to be. Einstein's 'correction' to Schrödinger was the Schrödinger eigenvalue equation.

3.2.5 FIAQ

I was told the quantum wave function is unobservable because it is complex. Can you observe both real parts of the complex electron wave? Good question! The answer is YES, which is why your books says NO. What is 'observable' in quantum mechanics was made early into an embarrassing mess. About the job of interpreting (or *helpfully presenting*) quantum mechanics, Murray Gell-Mann has written [9].

> 'The fact that an adequate philosophical presentation has been so long delayed is no doubt caused by the fact Niels Bohr brainwashed a whole generation of theorists into thinking that the job was done 50 years ago.'

Someone may be offended (but we don't know why) to hear that Bohr's influence and angst 75 years later has lost its relevance.

As we mentioned, Bohr, Heisenberg, and to some extent Born consistently maintained that quantum mechanics was *not supposed to be understandable*, and the wave function was *never observable*, because that premise fit their program. (THIS was not helpful.) Once the 'obvious un-observability' of complex numbers (wrongly) got a free pass, they dwelled on certain ultra-restrictive idealized measurements chosen for the word 'observables', which by construction supported their beliefs[19]. Meanwhile, we have no record of those gentlemen actually counting how many real numbers experiments could measure, compared to the number needed to specify a quantum state. In 1957 Ugo Fano showed [10] that there is always an equal number of *physically observable numbers* as needed to completely determine any quantum state. Fano's work actually goes beyond wave functions. It includes them, as well as the generalization called density matrices, discovered after wave functions. The number of variables to specify a state is finite in many cases: the *dimension* of the wave function or state depends on the model. There do exist *special cases* of some interest where the experimental conditions prevent measuring everything, but *that is not a universal rule.*

The embarrassment is that this took so long. Bohr and Heisenberg had made up their minds about a philosophy of unreality *before the actual quantum theory existed.* The philosophy was set in stone and maintained after Schrödinger discovered the

[18] Letter of Einstein to Schrödinger of 16 April 1926, in [8].

[19] The idealization of measurement is associated with the 'eigenvaiue postulate' it is intelligent to postpone, as we are doing. Understanding math facts that are trivial when you can see they are trivial greatly decreases the impression they have physical content, and vice versa.

previous physics was wrong. Physics waited 30 years for Fano to overcome Bohr's influence, and 40 years after that for Fano to get noticed in the mid-1990s. Isn't that amazing! With revitalized interest in *quantum computing*, the Fano process, now called *quantum tomography*, exploits the fact that the wave function is observable in principle. (More about this in section 8.2.3.) Measuring an infinite amount of detail may need an infinite number of experiments... but that was never the central issue, since one never needs infinite detail.

What was written above has contradictions. It's absolutely known that you cannot measure both the momentum and the position of a quantum particle. The authors who invented and made so much of the term 'quantum particle' defined their own word usage. Since there's no *parcle* entity in quantum theory, they were free to stick words onto quite inappropriate usages. When a plane wave is called a quantum *parcles*—just as those people did—it happens to have no definite position. This is just as obvious as it appears to be.

One does not expect tricky and narrow re-definitions of terms to create false conflicts in physics. Since physics words are tools without pre-determined meaning, it is very common for physicists to show no interest in word choices. It's unfortunate that word meanings then came to be so abused in quantum mechanics. *We didn't do it!*

Who controlled the use of words, and who benefited? The meanings were chosen to fit the particular order of presentation, putting point-*parcle*s *first.* Usually physics experts and teachers don't notice. Many had *unconsciously adopted a code-word system* taking for granted the premises of the *OQT*. Physics wars were fought over the right to claim the word 'principle'. The uncertainty *relation* turns out to be a math identity with no predictive physical information when it is derived from waves, or math identities. When the relation is called a 'principle', and presented as a claim about mysterious *parcles*, and well before a plane wave has been defined, it was made to appear to have content. Isn't that interesting?

Also, did you notice that physicists stopped spouting new 'principles' close to the time quantum mechanics was discovered. Far more physics has been found *after* basic quantum mechanics than all previous history, because quantum mechanics was the breakthrough allowing it to happen. Yet people had realized there were already too many bogus 'principles' on the market to give the word any market value.

As Feynman must have said:

It's not a great accomplishment to deal with something you can't explain by a Principle you can't explain. It is a great accomplishment to get rid of an unexplained Principle every time you can.

References

[1] Patterson L D 1949 *Isis* **40** 327

[2] Patterson L D 1950 *Isis* **41** 32

[3] Jardine L 2004 *The Curious Life of Robert Hooke* (New York: Harper-Collins)

[4] Nauenberg M 2006 *Robert Hooke's seminal contribution to orbital dynamics, Tercentennial Studies* (Farnham: Ashgate Publishing) pp 17–9
[5] Buckingham E 1914 *Phys. Rev.* **4** 345
[6] Thornton S T and Marion J B 1995 Classical Mechanics of Particles and Systems (New York: Academic)
[7] Goldstein H 1980 *Classical Mechanics* (San Francisco: Addison-Wesley)
[8] Przibram K (ed) 1967 *Letters on Wave Mechanics: Correspondence with Schrödinger, Einstein, Lorentz, and Planck* (New York: Philosophical Library)
[9] Gell-Mann M 1979 What are the building blocks of matter? *The Nature of the Physical Universe: 1976 Nobel Conference* ed D Huff and O Prewett (New York: Wiley) pp 29
[10] Fano U 1957 *Rev. Mod. Phys.* **29** 74

IOP Concise Physics

How to Understand Quantum Mechanics

John P Ralston

Chapter 4

Matter waves

Figure 4.1. As with 19th Century instruments, brass was the ideal material for the endcap calorimeter of the CMS detector at the Large Hadron Collider. With international cooperation, the Russian Navy allocated 600 tons of obsolete WWII battleship cartridges, which were melted down and rolled into 50 mm thick absorber plates.

4.1 Your quantum governmental representative

If you meet a US congressional representative, or a member of the House of Lords, be sure to tell him or her that everything in this civilization newer than animal power, wooden carts, and square-rigged ships is applied physics. All of the stuff the rich and powerful take for granted was once a shabby little demonstration physics experiment. The biologists, geologists, chemists, engineers, and mathematicians won't be insulted by this. We were all once the same people.

doi:10.1088/978-1-6817-4226-7ch4

Everything newer than 1900 is applied *quantum physics.* Without developing quantum physics, we'd have anthracite and bituminous coal, steel mills, the internal combustion engine, large electric motors we barely understood, and vacuum tube amplifiers nobody understood for our enormous radio sets. There would be one computer in each major country, occupying a square mile and burning up the resources of a city, and it would not have the capability of your smartphone.

A laptop computer screen violates several laws of physics of 1900. It produces light with almost no heat. It is a quantum device. The concept of a time-dependent display made of $1440 \times 900 = 1\,296\,000$ LED pixel triads (for three colors) was never dreamed by the wildest visionary of 1900. Anything with 4 million light sources would need 4 MW minimum to operate, and an army of men could not keep up with replacing the burnt-out bulbs.

The first visible light LED was invented in 1962 by Nick Holonyak Jr. Now they come in any color you desire. Before they manufacture LEDs the physics wizards compute a crystalline continuum where electron waves will do what they are told. While the calculations are cursorily *described* in terms of ad-hoc 'energy levels', the calculations actually engineered a host of *wave properties* from start to finish. The correct physical picture is not that electrons fall off shelves and spit out blips of light. The picture is much more like engineering Niagara Falls so that hoards of well-mannered electron waves maintain enough coherence for their *electromagnetic current* to make a natural laser-river of light. The current generates the light by

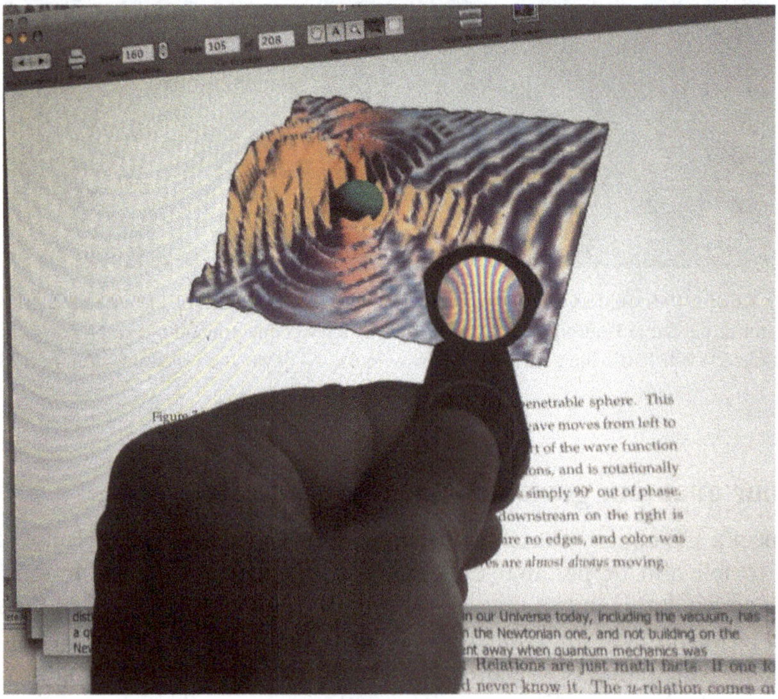

Figure 4.2. Pixels!

Figure 4.3. A good galena crystal (big cube) from the defunct Sweetwater Mine in southwest Missouri. The crystal escaped being broken up to make dozens of crystal radios. Only collectors care about them now.

the ordinary Maxwell equations, not quantum magic. The *extra* feature of quantum light explains why it's do darned coherent and loss-free. We recommend finding a mounted LED (plastic capsule, two leads) and exploring it with a 5–10 power hand lens. Many but not all basic LEDs are built with internals so they work by putting the leads across a few hearing-aid type batteries. Seeing the flow of light blazing from of a dust-mote crystal is one of life's great experiences[1]. A scientist of 1900 would not believe it was possible.

A scientist of 1900 would be very interested in a particle accelerator. Like the public, he'd assume the term was self-explanatory. He'd assume the device was made of brass[2] and wood and sat in the basement of an academic building. He would know about electromagnets, but not be a fool to believe in superconducting electromagnets. When you told him the ring was 27 km in circumference, the scientist would politely correct you: 'You mean 27 centimeters in circumference,' he would say, 'I have some experience with scientific instruments.'

4.2 A quantum device

Figure 4.4 shows a 'cat-whisker crystal radio'. The example is an elegant, artsy modern version of stuff kids could cobble together before quantum mechanics was

[1] This might be an exaggeration, but since it's quantum mechanics, we're not sure.

[2] Actually the CMS detector at the Large Hadron Collider accelerator has a small amount of brass in it: 600 tons.

Figure 4.4. A handsomely made crystal radio, a quantum device capable of picking up radio transmission without batteries or any other power source. Basic models are made by wrapping wire on a cardboard tube, and fiddling by hand with a fine 'cat whisker' wire on a galena crystal, shown in the inset. Radio designed and built by Richard H Weber.

on the scene. The device needed no batteries, no electrical power. That is amazing. Children (as well as stranded soldiers, seamen, etc) could hear radio transmissions from hundreds of miles with low technology. It was a quantum device, and yet *not a person on the planet understood it* in 1900, 1905, 1910, ... All that's needed is a coil of wire, wrapped to certain dimensions, a galena crystal, and a good ground[3]. For reasons not understood at the time, a very fine wire hooked into the system—the 'cat-whisker'—and poked about on any small bit of galena crystal[4] would let anyone hear tinny music in an earphone, or the address of the President of the United States, if the man[5] happened to be talking.

The crystal radio was a pre-quantum quantum device. Electron waves could go in one direction in the galena crystal from the cat-whisker, but not the reverse direction. The junction made a one-way electrical valve, called a diode, which is the crucial element to extract energy from the resonant circuit of the radio coils. The technicians of the day could tell you *what* the galena crystal did, but they had no conception of *why or how* it worked. When odd things happened there was no way to understand it. In 1907 H J Round noticed something unusual while experimenting with a cat whisker wire on a crystal of silicon carbide. He wrote:

[3] A 'ground' is a wire attached to some metal attached to the Earth. The Earth is a nearly infinite capacitor, allowing electricity to be dumped lavishly with minimal effects.

[4] Stranded soldiers not outfitted with galena crystals used razor blades touching pencil lead. Quite a few semiconductors will work.

[5] As long as they are all men, citing the US President as a 'man' is a gender-neutral insult.

> 'To the Editors of Electrical World: SIRS: During an investigation of the unsymmetrical passage of current through a contact of carborundum and other substances a curious phenomenon was noted. On applying a potential of 10 volts between two points on a crystal of carborundum, the crystal gave out a yellowish light. Only one or two specimens could be found which gave a bright glow on such a low voltage, but with 110 volts a large number could be found to glow .... In a single crystal, if contact is made near the center with the negative pole, and the positive pole is put in contact at any other place, only one section of the crystal will glow and that same section wherever the positive pole is placed.'

Round had discovered a natural LED but did not know what he had discovered. Shortly after 1900 not a single person in the human universe had any idea about how most of the world worked. Crystal radios, magnets, matter, whatever, almost nothing made sense.

As we know, Newtonian physics was the barrier. Electricity and electrical phenomena were the way out. Many of the breakthroughs of the 19th Century were made by *inventors*, not academic scientists. In 1832 Joseph Henry (who was an academic) observed *radio waves* when he noticed a spark sent 200 ft between coils, but did not follow up. In the late 1860s inventor Mahlon Loomis sent sparks between the Blue Ridge Mountains of Virginia, and sought support from the US Congress to develop a telegraph system [1]. Loomis's US Patent 129171 of 1872 includes:

> 'The nature of my invention or discovery consists, in general terms, of utilizing natural electricity and establishing an electrical current or circuit for telegraphic and other purposes without the aid of wires, artificial batteries, or cables to form such electrical circuit, and yet communicate from one continent of the globe to another .... I now dispense with both wires, using the Earth as one-half the circuit and the continuous electrical element far above the Earth's surface for the other part of the circuit.'

At least four bills were considered by Congress: one provided for 2 million dollars to incorporate the Loomis Aerial Telegraphic Company. In January 1873 a different bill was passed and signed into law by President U S Grant... but with no funds provided. In 1875 Thomas Edison was involved in a controversy about an 'etheric force' that communicated electricity across distances. Edison thought it communicated with the spirit world. Not until 1878 did George Fitzgerald (of Trinity College, Dublin) predict that electric circuits could produce electromagnetic waves, according to Maxwell's equations. (Maxwell himself never suggested it.) Despite the advantage of communicating with spirits, Edison despised formal education and favored DC electricity. In a dramatic industrial shoot-out associated with the Columbian exposition of 1893, Edison was soundly whipped by Nicola Tesla, whose training helped him understand AC power. Tesla is quoted for saying:

> 'If you want to find the secrets of the Universe, think in terms of energy, frequency and vibration.'

Figure 4.5. Nicola Tesla around 1890.

4.3 Electricity is a quantum effect

If Newtonian physics were your guide, you would never predict, and never understand electricity.

A lot about electricity contradicts the classical physics used to explain it. Everyone knows that the electric force field of an isolated charge decreases like the inverse radius-squared to the source. The force decreases so fast that waggling a reasonable charge at one end of a room produces effects almost too small to measure on the other side. Yet it's taken for granted that if you apply an electric field to one end of a metal wire, it appears at the other end with practically equal strength. Exactly how do electric field lines decide to flow along a metal tube, which is so convenient, without spreading out? It is 'explained' by a tautology: 'the surface of a conductor is an equipotential'.

Ohm's law says the decrease in electric potential is proportional to the current caused to flow. Those things called 'conductors' have an absolutely uniform electric potential over their entire length, so long as the current is zero or small enough. That's the statement: how does it occur?

Somewhat like the old quantum theory, the explanation of electricity in classical terms does not bother being complete. It is a mythology that's good enough to describe what's observed, *when applied just where the description is made*, yet it is not predictive and not internally consistent. Unlike the old quantum theory, nobody covered up the bugs of pre-quantum electricity and made them into philosophical features.

The 1900 *Drude model* is the default classical framework for electricity, which like most engineering compromises, seems to work well for cases where it applies. (Note that is a bit circular.) The Drude model is all about dissipation. Its myth proposes an ideal gas of point-like electrons in free space between atoms, getting accelerated by the electric field somehow oriented along the wire, and periodically having collisions with stationary atoms. As always presented, the model predicts the electric current $\vec{j}$ in terms of the electric field $\vec{E}$ by the rule

$$\vec{j}(\vec{x}) = \sigma\vec{E}(\vec{x}); \qquad \sigma = \left(\frac{n_e q^2 \tau}{m}\right).$$

The first equation is Ohm's law in local form. (There will be more about the local current in the next section.) The second equation defines the conductivity σ in terms of the electron number density n_e, the electric charge e_N in *MKS* units, and an average time between collisions τ. We noticed that Newtonian physics could not predict a fundamental frequency, so how does it predict τ, an inverse frequency? It does not! Drude's derivation cuts some corners, and τ is another fudge factor: the conductivity σ is observed in experiments, and converted to parameter τ by inverting the formula. Pretty cute, while testing *nothing*.

A priori, the actual prediction of the model would be that the conductivities of all substances would be about the same. That's because the chemical valence finds a 'few' possible free electrons per atom, and atoms do not differ wildly in their diameters. Multiplying factors of two or so all ways, most substances (inert gases excepted) should have conductivities varying by a factor of 10 or so. The opposite is observed: conductivities range from[6] 10^{-7} W m^{-1} to 10^{20} W m^{-1}. The collision time τ would range over 27 orders of magnitude. The Drude model works when it's circular and fails otherwise, and has become a mainstream tool of experimental interpretation. (Especially for experiments that 'don't depend on any theory'.)

The author's father was an electrician working under IBEW local 401 for decades. He was intelligent, loved music, chess, and math, and had a concise picture of electricity rather more complete than the Drude model. In the Ralston model electrons are like steel ball-bearings stacked up continuously along a long pipe. You hit a ball-bearing at one end, and its partner at the other end gets the impulse almost instantaneously. The virtue of the model is that the electrons are connected by pressure waves—waves!—and not inherently subject to 'friction', so that dissipation of energy (which is arbitrarily small, in an arbitrarily good conductor) is a side issue.

The quantum picture of electricity has the same elements, and explains everything. Electron waves flow smoothly across and inside the atoms of a crystal lattice, which are mostly electron waves to begin with. If the lattice is perfect the waves self-configure to flow perfectly, without dissipation, because they are flowing through themselves in perfect order. Waves of impulses are passed between and among electron waves because they are 'stiff' and interact. Unlike water, the wave equation for quantum electrons does not have a 'friction' term representing dissipation. In fact the flow of electrons is so frictionless it's a challenge to find where the electrical friction of resistance comes from. Resistance is dominated by *defects* in the crystal lattice, which disrupt the flow of electron waves like a boulder in a river.

[6] http://www.virginia.edu/bohr/mse209/chapter19.htm.

What about Earnshaw's theorem? Quantum mechanics does not contradict Gauss' law, so it still might seem a problem. It's all different with electrons as waves. There's no particle sitting at a point: waves are smeared out everywhere. There is a notion of *force*, but it's not captured by 'the force on the electron' because the *forces* are infinitesimal continuum wave stresses distributed everywhere. (Later we'll be more specific.) There are also forces Earnshaw knew nothing about. An important force is in the springiness of the electron wave, which does not want to bend. In fact, the potential energy per volume of an electron wave ψ has a term going like $|\vec{\nabla}\psi|^2$ or $-\psi^* \vec{\nabla}^2 \psi$. (Either expression gives the same result, added over the full volume.) The potential energy per volume of a static electromagnetic field has a term going like $|\vec{\nabla}\phi|^2$ or $-\phi^2 \vec{\nabla}\phi$. These expressions coincide in functional form, but come with different pre-factors, because they are generic features of the internal stresses of waves. So the quantum theory of matter has enough forces to balance out, and (very important) does not need to be static to produce a stable, steady Universe. The atomic electrons have the extra feature that they are constantly bouncing, gyrating, interpenetrating, and redistributing external stresses over an entire body such as a crystal. The electron waves typically merge into each other and share co-mingled existence for thousands of atomic lengths. That explains the fantastic strength of materials. Without that, cleaving a single layer of atoms 'glued' together classically would break metals into crumbs.

Earnshaw's theorem does not apply to a 'quantum force'. A piece of string does not exist classically, so it can transmit quantum forces. Consider a strange fact[7] of quantum strings[8]. You tie an ordinary string to a paperweight, and pull it a few inches across a desk. The atoms of the string and the paperweight come toward you. The *energy* you impart goes exactly the opposite direction, *from* your hand *into the object* to be dissipated by friction. The energy flows 'upstream' against the movement of atoms. Isn't that strange! It is not always this way. You can push the paperweight with a quantum pencil, and the energy will flow in the same direction as the atoms transmitting it.

4.4 The continuity equation

Quantum physics has a number of quantities that are not created or destroyed, and which are *locally conserved.* This is another exact feature of the theory, not a statistical one[9]. *Local* conservation means that when a quantity changes inside a volume, it must flow out the surface boundaries, just as water must flow through the

[7] The parable is borrowed from a more industrial version in [2].

[8] The 'quantum strings' getting a lot of promotion are not as revolutionary as the ordinary string. They are generally variations on models assuming quantum mechanics as a foundation, if any.

[9] It's annoying that petty scholars of schoolbooks tend to write that 'everything in quantum mechanics is statistical', adding 'at least when you measure it'. Try to notice that experimental measurement has statistical features regardless of the theory. It's a good idea to give no attention to claims for quantum measurements until you can check the assumptions and calculations.

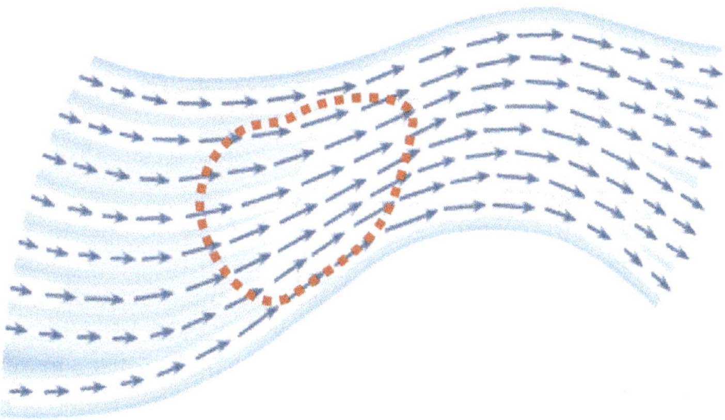

Figure 4.6. Quantum wave equations obey *continuity equations* prescribing flows of conserved quantities. For something conserved to leave a volume, it must pass through the surface surrounding it.

sides of a burlap bag to leak out; see figure 4.6. The conservation of stuff in quantum mechanics does not occur with discontinuous jumps or the exchange of *parcles*, real or virtual: those terms are a transcription of the smooth continuum equations into a hold-over language some think the popular press wants to hear. (*OK,* there are some weird things in quantum mechanics, but they can wait.)

The word 'local' means in the vicinity of an arbitrarily small spatial region. The local flow of things is described by a local *current*, symbol $\vec{j}\,(\vec{x},\ t)$, which represents the stuff passing at a point per oriented area per time. The *total* current is a different concept that has no information on position, but which is the local current summed over the whole area of flow. The total electric current in a wire has *MKS* units of amps: it is not specific about where inside the wire the current is distributed. As a rule, totals of physical quantities may appear in engineering rules, but they are not specific enough to help discover or understand physics in the continuum.

The *divergence* of the local current, symbol $\vec{\nabla}\cdot\vec{j}\,(\vec{x},\ t)$, measures its expansion or contraction point by point. An incompressible flow has zero divergence, which is exceptional. The *density* of stuff per volume $\rho(\vec{x},\ t)$ is closely related. If the density decreases in time, as measured by $\partial\rho/\partial t < 0$, the divergence will register an expanding current, $\vec{\nabla}\cdot\vec{j}\,(\vec{x},\ t) > 0$. Conversely if a current is converging with $\vec{\nabla}\cdot\vec{j}\,(\vec{x},\ t) < 0$ the density will locally be increasing, $\partial\rho/\partial t > 0$. The constraint expressing local conservation is the *continuity equation*:

$$\vec{\nabla}\cdot\vec{j} + \frac{\partial\rho}{\partial t} = 0 \tag{4.1}$$

While the equation has a time-derivative it is not an equation of motion. The continuity equation does not predict dynamics, but occurs when and if dynamical equations have certain *symmetries* that lead to local conservation laws.

Everything correctly reported as 'not created or destroyed' in physics is actually locally conserved. This includes[10] energy, momentum, angular momentum ... isospin, lepton number, etc, *when they are properly defined in all detail.* The conserved attributes of quantum waves are found from other formulas of currents distributed continuously over space. We cannot list everything here, because it is tantamount to listing all the equations of motion[11].

It follows that electric charge is a locally conserved quantum mechanical quantity. In the quantum theory the current is not a 'little dot' nor an independent substance but a locally conserved *feature* of charged quantum waves. *Given* a wave function, a formula computes the current it predicts. The local conservation law shuts the door on the loophole of a quantum electron disappearing in one place, and reappearing someplace else, with the total charge of the world conserved. This is not a minor point. Popular stories about 'non-local' quantum events correlated over distances tend to obscure the exact and unvarying facts of local conservation. Tricky suggestions of 'teleportation' might be interesting, but even *Star Trek* has the physics correct, so that matter going from one place to another needs to go through the space in between[12].

The quantum formula for the current (equation (6.1)) is not obvious, and perhaps impossible to guess without the Schrödinger equation. However the formula for the quantum current is the *only* new feature of the electromagnetic interaction. Maxwell's equations do not change with quantum mechanics. Let's take a moment to review this: in *MKS* units the current appears in

$$\vec{\nabla}\times\overrightarrow{B}=\mu_0\vec{j}+\mu_0\varepsilon_0\frac{\partial\overrightarrow{E}}{\partial t}$$

where $\vec{B}$ is the magnetic field. It is always possible to write[13] $\vec{E}=-\partial\vec{A}/\partial t$ and $\vec{B}=\vec{\nabla}\times\vec{A}$, where $\vec{A}$ is the vector potential. Computing $\vec{\nabla}\times\vec{B}=\vec{\nabla}\times(\vec{\nabla}\times\vec{A})$ gives

$$-\vec{\nabla}^2\overrightarrow{A}+\vec{\nabla}(\vec{\nabla}\cdot\overrightarrow{A})+\frac{1}{c^2}\frac{\partial^2\overrightarrow{A}}{\partial t^2}=\mu_0\vec{j}\,(\vec{x},t);\qquad \text{where}\quad c^2=\frac{1}{\epsilon_0\mu_0}. \tag{4.2}$$

[10] *Mass* is not a conserved quantity in quantum physics. As we have been learning, it is a *frequency parameter* for quantum waves. In an approximation neglecting interactions the frequency parameters add. The fundamental waves have conservation laws that lead to the appearance of a mass conservation law in Newtonian physics.

[11] Given Lagrangian $L(\phi,\partial_\mu\phi)$ for continuum mechanics, the Euler–Lagrange equations, $\partial^\mu(\partial L/\partial(\partial\phi/\partial x_\mu))=\partial L/\partial\phi$ are equivalent to continuity equations with sources, although seldom interpreted that way.

[12] 'Captain, the shields are up, we cannae use the transporter.'

[13] When $\vec{E}=-\partial\vec{A}/\partial t$ then Faraday's law $\vec{\nabla}\times\vec{E}=-\partial\vec{B}/\partial t$ is automatically solved. The option to express $\vec{E}=-\partial\vec{A}'/\partial t-\vec{\nabla}\phi$ defines a symbol $\vec{A}'$ in terms of the symbol ϕ. A common predilection defining ϕ to equal the Coulomb relation, or other conventions, makes this more complicated than it needs to be.

We've just computed the speed of light. The rest may appear intimidating, so pause to discuss some special cases:

- Suppose $\vec{A}$ is steady in time, and $\vec{\nabla}\cdot\vec{A} = 0$. Then $\vec{\nabla}^2\vec{A} \sim \vec{j}$. This is three copies of the Poisson equation, whose solutions from electrostatics are not too intimidating. Taking the divergence of both sides needs $\vec{\nabla}\cdot\vec{j} = 0$, which has an obvious meaning from the continuity equation.
- Suppose $\vec{\nabla}^2\vec{A} = 0$, which is just three copies of the Laplace equation. Then $\partial^2\vec{A}/\partial t^2 \sim \vec{j}$. This is an *acceleration* equation for $\vec{A}$, with $\vec{j}$ acting like the *force*.
- If $\vec{j}$ varies periodically, it will accelerate $\vec{A}$ periodically. The complete left side of equation (4.2) is a 'plain ordinary' wave equation, as if equation (3.1) were copied for each component of $\vec{A}$, but with an extra term $\vec{\nabla}(\vec{\nabla}\cdot\vec{A})$. That term kills the propagation of a longitudinal mode, so that light in free space is a transverse wave.

Since it's a wave the electromagnetic (EM) field can pass through regions ('tunnel'), refract, and bend ('interact'), and become trapped in sufficiently well-matched conditions. It is useful to compare light waves with matter waves, and also recognize the differences. First, the EM field can be created and destroyed. It has no conserved charge. Next, the EM wave equation has no intrinsic frequency parameter. It is pure and dimension-free, getting all information about the time and frequency scales of 'the rest of the Universe' from the current on the right hand side. Finally, there is a significant difference between the classical field $\vec{A}$ and the 'Schrödinger wave function' of the EM field. The classical EM wave is quite an exotic quantum state, where the fields are coherently correlated over gazillions of elementary excitations. (It is *not* a stream of statistically independent 'photons'.) In contrast the Schrödinger matter wave function at the basic level is for experiments concerning one electron, not many. The generalization to many electrons exists and will be treated in chapter 10.

4.5 FIAQ

You say you're a particle physicist, yet there are no particles. Then what do Feynman diagrams show? We wish the TV-spokespeople and physa-bloggers would tell you. Feynman diagrams are not pictures of reality. They represent mathematical tools to make approximate calculations.

Look at figure 4.7, which shows a diagram for Compton scattering. A number of moving plane wave fronts covers the figure. They should completely fill the whole region, interpenetrating everywhere, but the artist used restraint. The calculation of Compton scattering begins with the incoming waves, and describes how their interpenetration over the whole volume generates outgoing waves. The calculation involves multi-dimensional integrations of all the waves over the volume again and again.

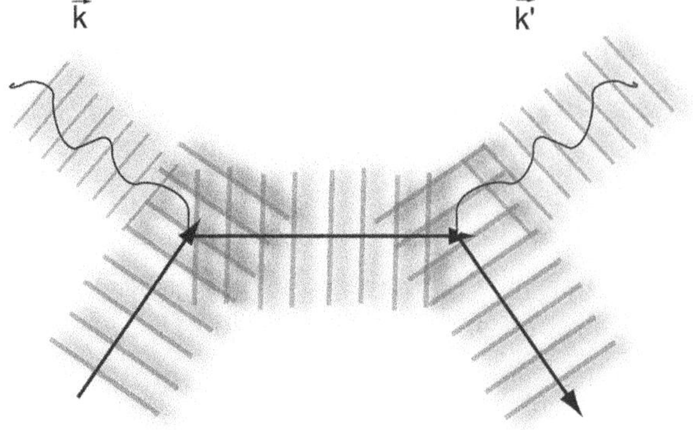

Figure 4.7. Feynman diagrams are not pictures of reality, but approximate calculational tools. The calculations involve plane wave fronts which are 'everywhere', even more extensively than shown. Removing the wave fronts and showing nothing but arrows and wiggles, representing wave numbers, creates a visual impression of 'particles'.

Mr Feynman had awesome calculational powers. He was not the first to make the multi-dimensional integrations, but he was the first to transcribe them into pictures. Instead of showing the wave fronts which are 'everywhere', Feynman amused himself by showing the wave vectors, and connecting them into pretended particle collisions. Erasing the wave fronts from the cartoon, they reappear anyway when you make the calculation. Some diagrams seem to show particles doing impossible things, like going in circles, or acting 'virtual'. Just as in tunneling, using *parcle* words and cartoons to replace what a wave does often produces an impossible interpretation.

There is something unnatural about the cartoons. Waves do not often configure themselves into rigid, infinitely extended plane waves. The calculations are too crude to do better! They use plane waves as an approximation, and do not attempt to be pictures of reality. Instead of solving smooth time evolution a few approximate waves generate more approximate waves, and so on. The rather crude nature of the calculations is sometimes recognized, with remarks that 'the exact process needs adding up an infinite number of diagrams for all particle exchanges'. Actually the approximations do not converge in any mathematical sense, but will eventually get *worse* if pushed too far. Feynman diagrams are great for their uses, but also reveal something interesting everyone should know: those diagrams of wave-exchanges are approximations. Nature is not making approximations. *Feynman diagrams are a charming, human baby picture of what nature is NOT doing.*

So how did 'particle physicists' get this wrong? In general, they don't! Trouble has been caused by changing word meanings that afterwards were misunderstood or misrepresented. By the time Klein and Nishina made calculation of *quantum* Compton scattering in 1929, the word 'plane wave' was starting to be replaced by 'particle' and vice versa. The replacement was convenient and unambiguous for those experts in the know. A definite *calculational* correspondence also made this easy. A plane wave is huge and complicated, but only has four parameters, which are the frequency and wave number $(\omega, \vec{k})$. It's the same number of symbols as the relativistic point *parcle* energy and momentum, $(E, \vec{p})$. It is not an accident, because the relativistic point *parcle* concept had been derived by observing little waves in the laboratory. There's nothing more to the issue, except for the *human error of extrapolating a little wave to a dimensionless point.*

References

[1] Gould H W 1988 Chronomogy of information and bibliography of Mahlon Loomis (1826–1886) http://www.math.wvu.edu/gould/Loomis%27Life.PDF

[2] Bridgeman P W 1943 *The Nature of Thermodynamics* section 34 (Cambridge, MA: Harvard University Press)

Chapter 5

More quantumy experiments

Figure 5.1. Waves are generally jumbled up, moving, interflowing, interpenetrating, and distributed everywhere over space. Waves do no select any 'position' concept. It is also OK to construct expressions for the average position of a wave-blob, along with other quantities. Photo by permission of Tommy Richardsen.

5.1 The Franck–Hertz particle accelerator

In two papers published in 1914, James Franck and Gustav Hertz reported experimental evidence for matter waves, but with only partial understanding. It is an example of how physics experiments can be partially understood, without seeing all the features. The timing of the experiment was ideal. It was 12 years before the wave theory of matter got developed. It was almost simultaneous with the botched and wrong pre-theory that appeared instead.

Franck and Hertz were experimenting with vacuum tubes, which means glass tubes with electricity going through low pressure gas. The gas was mercury vapor at 115 °C. This temperature was one where the vapor density was just right to see something interesting… empirical science depends on luck. As the gentlemen turned up the voltage, they expected to see more current, which is Ohm's law. At first the electric current passing through the tube went up, and then it dropped suddenly.

doi:10.1088/978-1-6817-4226-7ch5

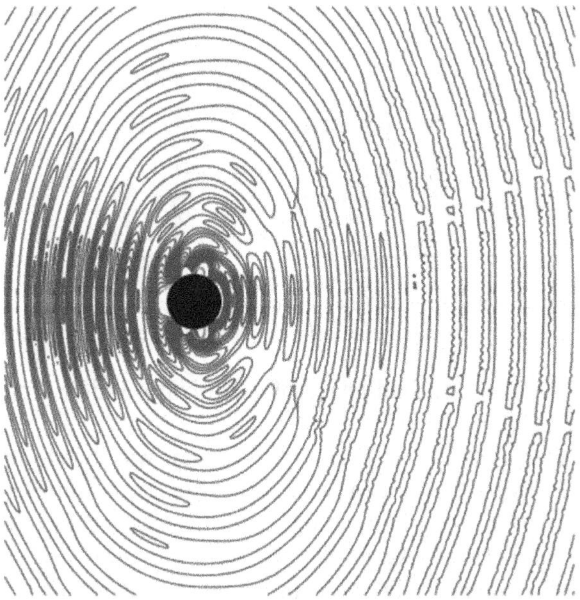

Figure 5.2. Scattering of a quantum wave moving left to right on a spherical obstacle. The three-dimensional shape is azimuthally symmetric about the horizontal axis through the middle.

Increasing the voltage further, the current increased from its minimum, and then dropped again suddenly. This cycle repeated every 4.9 V; see figure 5.4.

Franck and Hertz deduced that electrons with 4.9 V of energy[1] were stopping dramatically on the mercury atoms, and delivering practically all their energy. Hence, very little current. After being stopped once, electrons put under 2×4.9 V could accelerate again, and be stopped twice, and so on. A police 'speed-trap' model could explain this. When in free space electrons accelerate like dragsters, until they cross a limit, and the physics cops stop them. After paying the ticket, the electrons repeated the hurry-up-and wait cycle again and again.

The experimenters also observed increased light production near the current maxima. The most light comes out when the electrons stop, and deliver their energy to a 'new form of matter', emitting light upon decay. Not only that, but with some fussy adjustments, it is possible to see dark and bright layers in the vapor, clearly showing the zones where more collisions and light production is occurring. In the first paper they speculated the light was predominantly at wavelength 253.6 $m\mu$ (nm). That range of ultraviolet is not transmitted by glass. They remarked they would need a quartz tube and said: 'We hope soon to know whether in our case the emitted light belongs to the line 253.6 $m\mu$'. That was verified in the second paper. It was not trivial, because both glass and the human cornea block those wavelengths, making them invisible. The figure of 253.6 $m\mu$ (nowadays 2540 nm) came from measurements of mercury by the American R W Wood. Franck and Hertz were not

[1] One electron volt, symbol eV, is the energy gained by one electron charge traversing a change of potential by 1 V.

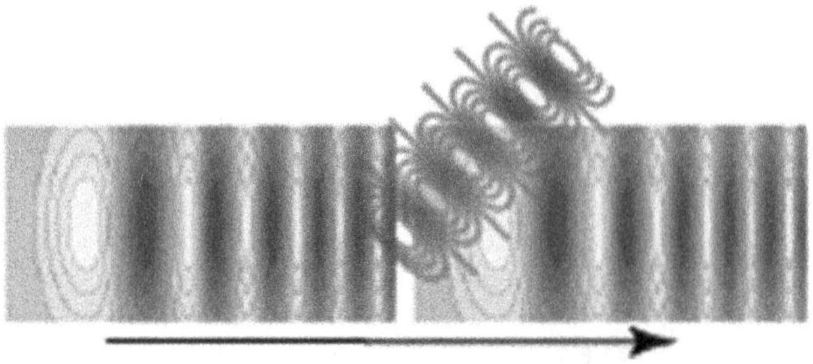

Figure 5.3. The first wave accelerator. In the Franck–Hertz experiment, electron waves picked up energy and frequency accelerating across a vacuum tube, periodically dumping the energy on resonant atomic transitions.

experimenting at random, but had heard of quantization ideas, for which they cited Stark and Sommerfeld.

Franck and Hertz initially thought the mercury was becoming ionized at the speed-trap energies. Ionization means electrons are kicked completely out of atoms. The dramatic effect of nothing happening, then a sharp energy loss, was consistent. Actually Franck and Hertz were playing with a *resonance* of the atom. To understand what's happening, imagine blowing air across a pan-pipe, or a glass bottle. Below a critical speed not much happens. When the air speed is perfect the pipe will resonate, extracting energy, and converting it into the natural vibrational frequency of the system. Each mercury atom is like a little pan-pipe, which is somewhat fussy to the passing electron wave, but then responds with enthusiasm when the frequencies are matched just right.

The resonance of a mercury atom is not really a new form of matter, but it is certainly a transformation of the electron waves in the atom from one set of vibrating shapes to a distinctly different one. The analogy suggests there ought to be many resonant frequencies: for a pan-pipe, those would be the harmonics. It's a bit too simplistic to expect *perfect multiples* of a given frequency… that may or may not happen. The Franck–Hertz experiment was a success partly by playing with *one* repeated resonance of cold atoms, avoiding other complications.

A fundamental thing about resonance is that *frequencies of the driver and resonator must match*. If Franck and Hertz had known about matter waves, and deduced that frequencies were matched, they would have jumped 15 years ahead on the research curve, and saved us a century of propaganda. Recall we estimated that $10^{16}s^{-1}$ is the approximate frequency of an electron wave making an atom. The Franck–Hertz experiment's light with wavelength $\lambda = 2536$ Å has a frequency $\nu = c/\lambda = 1.18 \times 10^{15}s^{-1}$. The numbers are close enough, and might be dismissed as just 'typical atomic physics'. However, the *matching of frequencies* implies that the electron wave frequency was $1.18 \times 10^{15}s^{-1}$. This re-calibrates the old fashioned unit called the 'volt'. The volt is a rather quaint unit which for years was based on electrochemical battery cells, which are reproducible, but not very informative. Franck and Hertz had the correct idea that 4.9 V measured the electron's motion. That motion is not a *parcle* but a wave phenomenon: the data says the extra frequency acquired by an electron pulled with 4.9 V is $1.18 \times 10^{15}s^{-1}$. In their *Zusammenfassung* (Summary) Franck and Hertz wrote: [1]

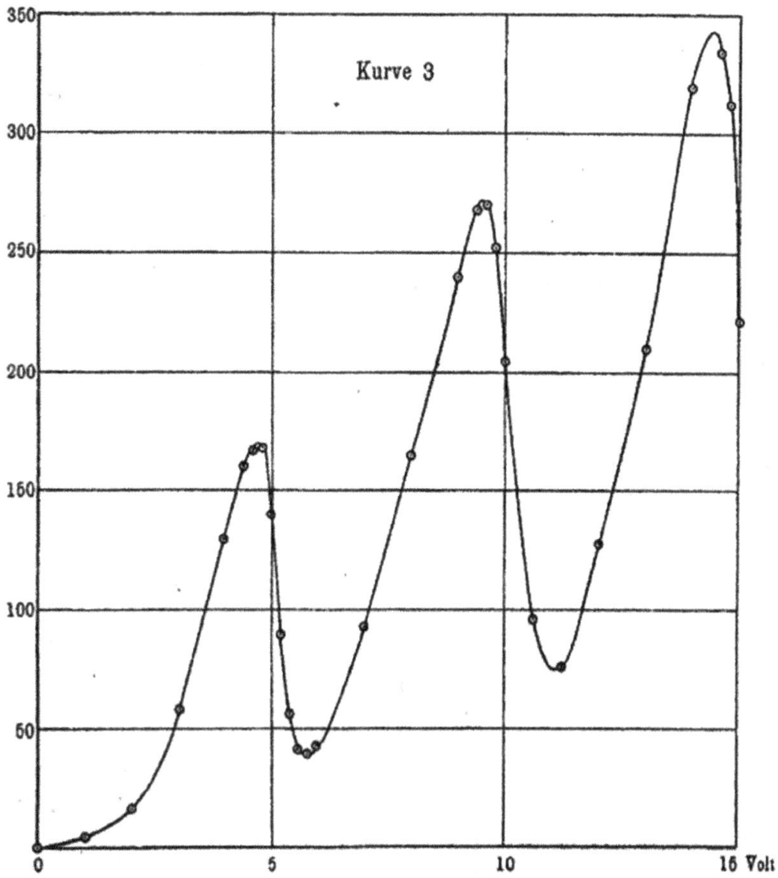

Figure 5.4. Figure 4 from the first 1914 paper of Fanck and Hertz. The curve shows current in the tube versus voltage along the x axis. Markers show 0, 5, 10, and 15 V.

'Es wird gezeigt, daß die Energie eines 4,9-Voltstrahles genau gleich einem Energiequantum der Quecksilberresonanz linie 253,6 $\mu\mu$ ist.'

(It is shown that the energy of a 4.9 V beam is exactly equal to an energy quantum of the mercury resonance line 253.6 $\mu\mu$.)

That gives the frequency–volt conversion factor in one step:

$$\begin{aligned} &4.9 \text{ electron volt} \rightarrow 1.18 \times 10^{15} \text{ Hz}; \\ &1 \text{ electron volt} = \frac{1.18}{4.9} \times 10^{15}\, Hz = 2.4 \times 10^{14}\, Hz. \end{aligned} \tag{5.1}$$

In 1962 Brian Josephson, a 22 year old student who could do basic quantum mechanics, predicted the relation between frequency and voltage that would occur in a superconducting junction device. The experimentally determined number is called the Josephson constant K_{J-90}, which has the value

$$K_{J-90} = 0.4835979\ \frac{\text{GHz}}{\mu\text{V}},$$
$$\text{or}\ \ 1_{J-90}\text{V} = 4.836 \times 10^{14}\ \ \text{Hz}.$$

This is exactly twice the value of equation (5.1). The J–90 volt *must* be twice the value as before, because superconductor current is made with cooper pairs, charge $2e$. The AC Josephson effect is *stupendously* simple. You put a DC voltage across a junction, and the current oscillates at 4.836×10^{14} Hz V^{-1}. It was a schoolbook calculation that had been missed by schoolbook calculations made for waves with *one single* frequency.

Let's check a previous result. It is known that hydrogen can be ionized by 13.6 V. That means you might put (say) 10 1.5 V AA batteries in series, accelerate electrons emerging from a heated filament, and watch them destroy hydrogen atoms that get in the way. A 13.6 V electron will have an incoming frequency $\nu_{13.6}$ we can compute:

$$\begin{aligned} \nu_{13.6V} &= 13.6\ \text{V} \times 2.4 \times 10^{14}\ \text{Hz V}^{-1} = 3.3 \times 10^{15}\ \text{Hz}; \\ \omega_{13.6V} &= 2\pi\nu_{13.6V} = 2.08 \times 10^{16}\text{s}^{-1} = \omega_H. \end{aligned} \tag{5.2}$$

This *predicts* the hydrogen Rydberg frequency ω_H perfectly, equation (2.1). There is one subtle thing: with equation (2.1), we mentioned that the frequencies of the electron wave current are the *differences* of the wave vibrational frequencies. If the incoming electron frequency is 13.6 V, to match the atomic resonance frequency and cause ionization, the *final* electron frequency must be zero, so the difference is 13.6. The final electron stops dead: an electron at rest is the only case where the wave can have zero frequency. The shape of such waves can be computed: they are flat, with no bending, no motion.

Comment. Notice we've compared data to data using the wave theory, without ever needing Planck's constant. It is never needed in quantum theory, while always needed in the old quantum theory. The critical element we bypassed was the Newtonian electron. It would have caused the following extra, and meaningless steps: • Convert the electrochemical electron volt to *MKS* Newtonian kinetic energy. • Assert that conservation of *MKS* energy applies to quantum Newtonian electrons, whatever that means. • Assert that atomic electrons make an instantaneous quantum leap between unexplained energy levels. • Assert that the emerging light will carry off the *MKS* energy. • Convert the *MKS* light energy back to frequency, which is what a spectrometer actually measures. Quantum mechanics does not give you cookie-cutter formulas for *MKS* Newtonian energy. Quantum mechanics destroyed Newtonian physics as a way of thinking about nature. (That still permits the *MKS* unit system to mark up experimentally observed quantities, such as the Josephson constant, with Newtonian symbols and $\hbar$. And Newtonian physics will always be around for engineering physics.)

Continuing. A modern particle accelerator is not that different. Electrons, protons, or whatever are electrically stripped from atoms. After acceleration they collide, and the experimenters measure the rate. Observing a significant change in the collision rate, at a particular collision energy, is enough to know a new resonance

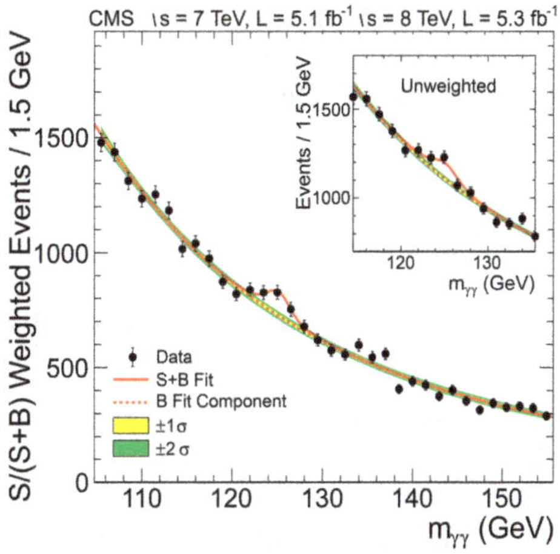

Figure 5.5. A histogram (empirical distribution) of counts of two photons versus invariant mass from an early CMS experiment at the CERN Large Hadron Collider. The invariant mass is computed from certain combinations of the wave numbers and frequency, and owes nothing to the archaic Newtonian definition. The little bump on the curve shows an excess production rate near the 125 GeV resonance point predicted for the 'Higgs particle'. It is no little dot, but a long-sought new vibrational dimensionality of the Universe, an independent quantum field.

has been seen. When the Higgs 'particle' was discovered, probably, at CERN in 2012, nobody saw any little dot with any kind of Newtonian mass. What was seen was a 'bump' in a chart, measuring an increased collision rate. It is rare to make a Higgs, and the overall collision rate (like the current of Franck–Hertz) changes by a relatively small amount. The visible signal of the Higgs resonance was an increased amount of light quanta tuned exactly to the 125 GeV resonant frequency. (Two photon waves of about 62.5 GeV each, plus Lorentz boost effects, if you want to be technical.)

This was a very difficult experiment with a very small signal. Previous experiments searched for years, but did not have the necessary technology. The public was told it was a great discovery, and it was, yet we in the physics business were also disappointed. The Higgs *field resonance* was exactly where theory using previous experiments had predicted it should be[2]. Yet the satisfaction of confirming a theory is nothing like the thrill of breaking one, which many had hoped for.

5.2 The Davisson–Germer demonstration experiment

With Schrödinger's equation a calculation of a few lines could predict the conditions to make a dramatic *demonstration experiment*, of the kind that would convince a dead man.

[2] The Higgs mass parameter, like all the others of the Standard Model, is not determined by the model. Yet if the theory was right, the Higgs mass parameter had essentially been observed already via the interplay of precise measurements and theory.

The opportunity to perform such an experiment was given to Clinton Davisson, who had the experience and resources at hand. That was going to be a Nobel Prize enterprise, including the press release. Davisson's Nobel lecture begins:

```
That streams of electrons possess the properties
of beams of waves was discovered early in 1927
in a large industrial laboratory in the midst of a great
city,
and in a small university laboratory overlooking
a cold and desolate sea.
```

Note the date where Davisson cites himself for what the theorists told him to go discover. The part that is true is that a cold desolate sea lies below the University of Aberdeen, where George Padgett Thomson and students did a simpler, independent experiment on a small budget at the same time. Eventually the dissertation advisors shared the 1937 Nobel Prize, and thanked their students.

Four pages into the lecture Davisson divulged more of the truth: 'The first to draw attention to it (electron diffraction) was Walter Elsasser who pointed out in 1925 that a demonstration of diffraction would establish the physical existence of electron waves.'. Walter Elsasser (1904–91) did more than make a casual suggestion. His paper [2] suggested two tests of de Broglie's wave proposal that perhaps had *already been observed.* One test was the Ramsauer (or Ramsauer–Townsend) effect, where electrons with about 1 eV of energy were observed to have practically zero collisions with noble gases such as argon. The electrons flowed past the obstacles as *waves* matched to the atomic size. Another test was the observation of peculiar ripples in the angular distribution of electrons scattering off platinum, as observed by Davisson and Kunsman in 1923. Elsasser showed by calculation that both phenomena were consistent with de Broglie matter waves. Elsasser proposed a more quantitative demonstration experiment, and sought the support of James Franck of Göttingen (the same Franck as Franck–Hertz). Franck was encouraging but would not spare manpower to help the 21 year old Elsasser, who was inexperienced and eventually gave up experimental physics as being too hard. Davisson knew of Elsasser's paper [3] but did not believe it. At the time Schrödinger had not yet found his wave equation, and the absence of a wave equation made analysis very difficult. After Schrödinger published his papers on wave mechanics, Born in 1927 remembered Elsasser, who happened to be his own student. Born presented the old Davisson–Kunsman data at an Oxford meeting as evidence for matter waves. Davisson attended the meeting, and after talking to important people like Born, Schrödinger, and Thomson, 'they all told him to look for wave interference'. So then he believed it. Later Elsasser said about Davisson:

> 'So by that time Schrödinger's work was known, you see, and then he (Davisson) just gave me a nice footnote and that was that. As a matter of fact, you see I was just 21 years old when I wrote this, and I was just too young to exploit it.'

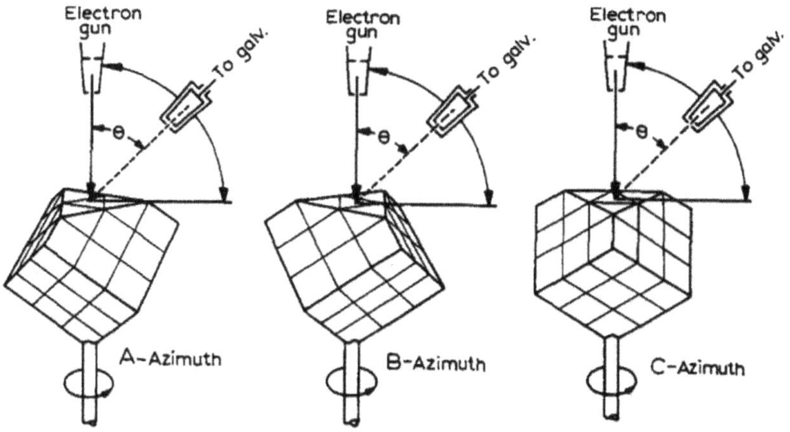

Figure 5.6. Schematic diagram of the Davisson–Germer experiment showing primary beam (vertical) and detector pickup ('to galv'). The blocky thing represents a tiny crystal of elemental nickel, which was rotated about the vertical axis to determine the angular distribution of scattering. From Davisson's Nobel lecture. © 1937 The Nobel Foundation.

Yet since Elsasser predicted this particular Higgs particle, and suggested the crucial experimental element of using a single crystal, we think he might have shared in the Nobel prize, despite not having 'a large industrial laboratory in the midst of a great city'. Elsasser went on later to contribute to nuclear physics and become a highly distinguished geophysicist.

Figure 5.6 shows a schematic of Davisson and Germer's experiment from the Nobel lecture. In 'good' directions the atomic planes of the target crystal caused waves to constructively interfere, increasing the scattered intensity. How to convert such a pattern into information about the impinging wavelength was quite well known from x-ray scattering.

5.2.1 Matter waves tend to be small

The measured intensity as a function of the azimuthal angle (around the rotation axis) is shown in figure 5.7. Since the atomic plane spacings were known from x-ray scattering, the wavelength $\lambda_{scatt} = 1.65 \times 10^{-8}$ cm fit the data. (This is $n = 1$ fit listed in their table 1.) A single clean wavelength, which is exceptional, occurred for the experimentally circular reason that no other wavelength (wave packet) was going to match the crystal lattice and electron beam energy. (That's how great experiments tend to work out.) The potential applied was 54 V. Converting to frequency gives $54\ \text{V} \times 2.4 \times 10^{14}\ \text{Hz V}^{-1} = 1.3 \times 10^{16}$ Hz. The data for frequency and wavelength is

$$\lambda = 1.65 \times 10^{-8}\text{cm}$$
$$\nu = 54\ \text{V} \times 2.4 \times 10^{14}\ \text{HzV}^{-1} = 1.3 \times 10^{16}\text{s}^{-1}.$$

Note the order of magnitude: the numbers are typical atomic sizes, one more time.

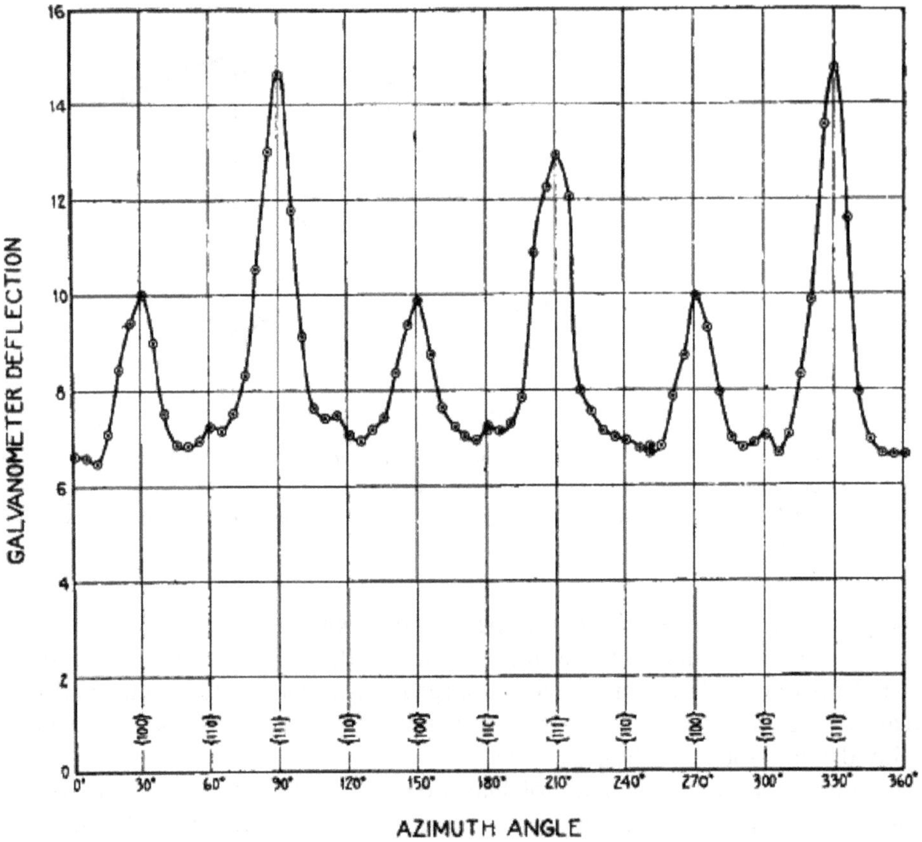

FIG. 2.—Intensity of electron scattering vs. azimuth angle—54 volts, co-latitude 50°.

Figure 5.7. The intensity of electrons observed by Davisson–Germer as a function of azimuthal angle, from their 1927 *Physical Review* paper. Bumps show maxima representing constructive interference at certain angles. The periodicity of the pattern is a direct consequence of the crystallographic symmetry.

5.3 The free space Schrödinger equation

In section 6.2 we will review how Schrödinger guessed his famous equation. Here we consider the special case of propagation in free space, or 'free quantum electrons'. The equation in free space is

$$\iota \frac{\partial \psi}{\partial t} = -\frac{1}{2\mu_e} \vec{\nabla}^2 \psi; \qquad \mu_e = \frac{\omega_e}{c^2}. \tag{5.3}$$

The complete equation adds $U(\vec{x})\psi$ to the right hand side, where U describes an interaction. Far from interacting regions one has $U(\vec{x}) \to 0$.

If you promise not to be bullied, nor get angry, the next step is a very ordinary expression which has scared or angered many. At time $t = 0$ you send out your sub-quantum minions to push and pull a matter wave into an arbitrary shape $\psi_{\text{any}}(\vec{x})$, and hold it there. FREEZE THE WAVE IN PLACE, and we seriously mean in *ANY* shape: no restrictions, because *the initial shape comes from arbitrary and undetermined initial conditions.* The mathematics of Fourier analysis can express any shape as a sum of complex cosine waves,

$$\psi_{\text{any}}(x) = \int dk\ e^{\imath kx}\tilde{\psi}_{\text{any}}(k); \leftarrow \text{no information!}$$
$$\psi_{\text{any}}(\vec{x}) = \int d^3k e^{\imath \vec{k}\cdot\vec{x}}\tilde{\psi}_{\text{any}}(\vec{k}) \leftarrow \text{no information!} \tag{5.4}$$

Here $d^3k = dk_x dk_y dk_z$. These expressions (one- and three-dimensional form) are not predictions and not puzzles to be computed, but more like water-color paintings waiting to be painted. Whatever $\psi_{\text{any}}(\vec{k})$ is put inside the integral, it will predict a corresponding $\psi_{\text{any}}(\vec{x})$, and vice versa. We have used *the trick of putting no information* into the expressions, except the fact there's two related ways to express *any* function.

Next, we tell the sub-Q minions to release the wave and let it slosh around. No matter what the time dependence may be, we can encode it replacing $\tilde{\psi}_{\text{any}}(\vec{k}) \to \tilde{\psi}_{\text{any}}(\vec{k},t) = \tilde{\psi}(\vec{k}, t)$. That's equivalent to using the no-information trick at each moment in time. The time dependence comes from physics, which the Schrödinger equation will predict. To avoid confusion we are dropping subscript-*any* because the time dependence is *not* arbitrary. Once again express the time dependence with a Fourier transform:

$$\psi(x, t) = \int dk d\omega\ e^{\imath kx}\tilde{\psi}(k, \omega)e^{\imath kx - \imath\omega t} \leftarrow \text{no information!} \tag{5.5}$$

Whatever $\tilde{\psi}(k, \omega)$ is, it will correspond to $\psi(x, t)$, and vice versa.

Now for physics: we now 'educate' or direct the meaningless expression above to solve the free Schrödinger equation:

$$\left(\imath\frac{\partial}{\partial t} + \frac{1}{2\mu_e}\frac{\partial^2}{\partial x^2}\right)\psi(x, t)0,$$
$$\int dk d\omega\ e^{\imath kx}\tilde{\psi}(k, \omega)\left(\omega - \frac{k^2}{2\mu_e}\right)e^{\imath kx - \imath\omega t}\to 0. \tag{5.6}$$

Inside the integral we used the *primeval eigenvalue equation* twice:

$$\imath\frac{\partial}{\partial t}e^{-\imath\omega t} = \omega e^{-\imath\omega t}$$
$$\imath\frac{\partial}{\partial x}e^{\imath kx} = ke^{\imath kx}. \tag{5.7}$$

Any time an operation on a function produces the same function multiplied by a constant, it is called an *eigenvalue relation.* The functions associated with eigenvalues are called *eigenstates.* The equations above are eigenvalue relations for derivative operators. It is not much at first, but the primeval equation is the tip of a very big mathematical iceberg that comes up soon.

The right hand side of equation (5.6) will be zero for all x, t if and only if the integrand $\psi(k, \omega)(\omega - k^2/2\mu_e) \to 0$. Whenever $\psi(k, \omega) \neq 0$ we must have $\omega - k^2/2\mu_e = 0$. But $\psi(k, \omega)$ was an arbitrary function of k (recall *any*), so all solutions must obey the *free space dispersion relation*

$$\omega = \frac{k^2}{2\mu_e}. \tag{5.8}$$

The range of wave numbers k is undetermined, and equivalent to the initial conditions, equation (5.4). For each k there must be a frequency ω given by equation (5.8). That is the content of the free space Schrödinger equation and it is the *only content*.

The expansion in Fourier modes worked beautifully because it was an *expansion in eigenstates* of the differential operators $\imath\partial/\partial t$ and $-\imath\partial/\partial x$. Each operator produced an eigenvalue. Finding eigenvalues of the frequency operator happens to be the general way to solve the Schrödinger equation.

Example: determining μ_e. With $\mu_e = 0.87$ scm^{-2}, and $k = 2\pi/\lambda$, the equation predicts for the Davisson–Germer experiment

$$\nu = \frac{\omega}{2\pi} = \frac{2\pi}{1.74\ \text{s}\ \lambda^2/\text{cm}^2} = \frac{2\pi}{1.74 \times 1.65^2 \times 10^{-16}}\text{s}^{-1} = 1.33 \times 10^{16}\ \text{Hz}, \tag{5.9}$$

just as reported by the data. Or you can reverse this: the data of Davisson and Germer predicts $\mu_e = 0.87$ scm^{-2}.

Example: a general solution. The double Fourier expansion with no information in it, equation (5.5), represents the solution when the dispersion relation is put into it, as follows:

$$\psi(x, t) = \int dk\ e^{\imath kx}\tilde{\psi}(k)e^{\imath kx - \imath k^2 t/2\mu_e} \leftarrow \text{general solution.} \tag{5.10}$$

In three dimensions replace $dk \to d^3k$ and $kx \to \vec{k}\cdot\vec{x}$. We suggest you check this by applying the differential operator $(\imath\partial/\partial t + \frac{1}{2\mu_e}\partial^2/\partial x^2)$. Once again $\tilde{\psi}(k) = \tilde{\psi}_{\text{any}}(k)$ is a code for the initial conditions. *That's all there is to it!* We have solved a quantum mechanics problem… nothing more is needed! Notice there is no integration $d\omega$ because $\omega \to k^2/2\mu_e$ is fixed inside the equation. The same result comes from integrating $d\omega\delta(\omega - k^2/2\mu_e)$, in case you are familiar with the Dirac delta function. In any event, the physics is predicting the bare minimum information, which is all that is needed.

Example: de Broglie. Louis de Broglie thought $E = h\nu$ and $p = h/\lambda$ were fundamental relations. They were assembled with $E = p^2/2m_N$ to give $h\nu = (h^2/2\lambda^2 m_N)$ for each 'quantum *parcle*'. The free space dispersion relation $\omega_e = k^2/2\mu_e$ is the same thing, but with a very different meaning: there is no single wave in the general solution. The only case where one would have a single wave with a single frequency and wavelength would be some fabulously precise selection as an initial condition. The Schrödinger equation predicts the de Broglie relation when reduced to that single case: the de Broglie relation *does not* predict the Schrödinger equation, because it was too narrowly posed from the start.

Example: wave speeds. The general solution is a sum of harmonic waves that are functions of $\vec{k}\cdot\vec{x}-\omega t$. A function of $x - vt$ translates rigidly to the right at speed v, as if drawn on clear plastic, and pulled across a page. Write

$$\vec{k}\cdot\vec{x}-\omega t = |\vec{k}|\left(\frac{\vec{k}}{|\vec{k}|}\cdot\vec{x}-\frac{\omega}{|\vec{k}|}t\right) = |\vec{k}|\left(\hat{k}\cdot\vec{x}-v_{phase}t\right),$$

$$\text{where}\quad v_{phase} = \frac{\omega_{\vec{k}}}{|\vec{k}|}\frac{\vec{k}^2}{2\mu_e|\vec{k}|} = \frac{|\vec{k}|}{2\mu_e}.$$

Each electron plane wave moves rigidly at the phase velocity $v_{phase} = |\vec{k}|/2\mu_e$. This is different from the cookie-cutter $c = \lambda f$ equation. The velocity of free space matter waves depends on $\vec{k}$: the larger $|\vec{k}|$, the smaller the wavelength, the faster they go. As we will see, the interactions that bend quantum waves is much like a local distortion of their propagation speed in the background 'medium'.

5.3.1 Interpreting the sign of the frequency

The free space wave shows once again how 'quantum mechanics' is a misnomer. The values of ω and $\vec{k}$ are undetermined, and completely continuous. One says there is a 'continuous spectrum' for waves 'propagating in the continuum' of free space. There is no 'quantization' to be found here. The discovery of the Schrödinger equation is that frequencies will be quantized when and if the dynamical solutions have quantized frequencies. In general, frequencies are *only quantized* when the waves are localized by an interaction that traps them. Such things are called 'bound states'.

The frequency of a matter wave is *positive* when propagating in the continuum and *negative* in a bound state. Consider figure 5.8 showing a bound state. In the interior region the wave has the strongest interactions, which will be complicated. Towards the outside boundary the interactions get small, and the wave attempts to propagate out into free space. In that region $\vec{\nabla}^2\psi$ is small, with a particular sign. The sign is best understood by dividing out ψ. In the figure for $x \gtrsim 2$ the value of $(\vec{\nabla}^2\psi)/\psi > 0$, causing the wave to flatten out as it approaches the boundary. For example $\psi(x) \sim e^{-kx}$ has $d^2\psi/dx^2 \sim k^2\psi(x)$, with $k^2 > 0$.

The time dependence of the wave $\partial\psi/\partial t$ must maintain that trapping. For each self-resonant trapped wave with time dependence $\exp(-\imath\omega t)$, the time derivative goes like

$$\imath\frac{\partial}{\partial t}e^{-\imath\omega t} = \omega e^{-\imath\omega t}.$$

If we are sufficiently far from the interaction center to neglect the interaction term, we have

$$\omega\psi \sim -\frac{k^2c^2}{2\omega_*}\psi, \tag{5.11}$$

which needs $\omega < 0$.

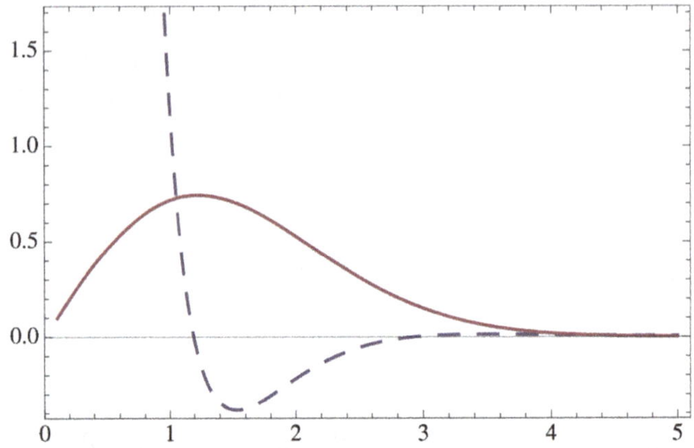

Figure 5.8. The frequency of a trapped eigenstate is negative ($\omega_n < 0$) which corresponds to ($\vec{\nabla}^2\psi)/\psi > 0$, solid curve) in regions near the outside edges of an atom, where the interaction function (dashed) becomes sufficiently small.

Next consider propagation in free space. A generic plane wave goes like $\exp(\imath kx)$, so that $\psi''(x) = -k^2\psi(x)$. Then equation (5.11) is replaced by $\omega\psi \sim -(-k^2)\psi$, needing $\omega > 0$.

Finally there is the delicately balanced case with $\omega = 0$ for the perfectly flat wave $\vec{\nabla}^2\psi = U\psi = 0$. The signs are summarized by

trapped wave, 'bound state' $\omega < 0$;
flat wave $\omega = 0$;
untrapped wave, 'unbound state' $\omega > 0$.

5.3.2 The memorized substitution rules

We presented the free space Schrödinger equation without explaining it, because it is more urgent to see how it works. We think explanations are important, and it will appear in section 6.2. However, the campaign promoting the *non-explanation* of the Schrödinger equation came early, and was very well advertised. It fits in pretty well with the idea a person was not supposed to understand quantum mechanics.

The main key to the not-understanding tradition was memorizing non-explanations. We're not going to defend it. In a typical Copenhagen presentation one will see statements like this:

- 'For every classical physical quantity, such as position, momentum, angular momentum, mass[3], etc, ... quantum mechanics has a Hermitian operator using the same symbol'. And this cannot be explained.

[3] *Liboff* explicitly includes *mass* as an operator, which it ain't. More than one versions of the memorized substitution rules make no sense.

- The *momentum* of the quantum particle is represented by $\vec{p} = -\imath\hbar\vec{\nabla}$. And this cannot be explained.
- For each system with a classical Hamiltonian $H(\vec{x}, \vec{p})$ there is a Hamiltonian operator $H(\vec{x}, \vec{p})$ replacing $\vec{p}$ the numbers by $\vec{p}$ the operator. This is called 'quantizing the system'.
- After quantizing the system, its stationary states are given by solving $H|\psi> = E|\psi>$.

The memorized substitution rules (*MSR*) include the above and a little more. In section 3.2.2 we mentioned Poisson brackets were assumed not to be understood by novice students. So the *MSR* say that for every classical Poisson bracket, one should use a corresponding commutator of operators. If this appears to be a *non-sequitur*—namely a statement not connected in a logical or clear way to anything said before it—that was the character of the *MSR*.

Many studying quantum mechanics have been perplexed. Actually students use the *MSR* as a way to remember formulas. The Newtonian formula for a Hamiltonian is $H = \vec{p}^2/2m + V_N(\vec{x})$, where V_N is the Newtonian potential energy. Substituting $\vec{p}$ gives $H^{\rightarrow} - \hbar^2\vec{\nabla}^2/2m_N + V_N(\vec{x})$. The last step finds the stationary states:

$$-\frac{\hbar^2\vec{\nabla}^2|\psi>}{2m_N} + V_N(\vec{x})|\psi> = E|\psi> .$$

The equation just written uses *terribly sloppy, misleading harmful notation* (see section 7.1.3), but *the whole memorized kludge* is sloppy, so what's not to like?

Many notice what's not to like. Not one thing has been explained, nor understood, by enacting memorized rules. Einstein objected to what became an inherently irrational acceptance of unthinking, which he called the 'Heisenberg–Bohr tranquilizing philosophy—or religion?' (And this was not the debate about quantum probability, either.) If you take a physics course and get points for 'quantizing' an arbitrary system—a favorite is the 'banana peel'—try to remember it's just a schoolbook algorithm to name differential equations. The only Hamiltonians that matter are those describing nature, not those predicted by recipes. In section 8.2.1 we will review the relation of $-\imath\vec{\nabla}$ to 'momentum' that is not an inexplicable postulate, but a *derived fact* of the Schrödinger equation.

5.3.3 Don't add the wave functions of two electrons

There's a joke about two electrons who go into a bar. The bartender says: 'What will you have?' One electron says: 'How much is a double-slit?' The bartender says: '$8.' The other electron says: 'Count me in, we'll both have the same.' The bartender says: 'That'll be $64, cash.' The electrons complain: 'Can't you add?.' The bartender says: 'You waves are all the same. And I know how you multiply.'

Oooh... When quantum subsystems are composed, their wave functions *multiply*, they do not *add*. In principle it is a *procedural decision* for wave functions to compose by multiplication, but it is the only consistent option. The consequences of products and *entanglement* are the topic of chapter 10. The way it works is totally unexpected, and was never guessed before quantum theory. We need to give a hint of it here, to avoid mistakes and paradoxes that might come up if you adventurously (and incorrectly) start superposing the waves of more than one quantum system.

Consider two frequency eigenstates called systems 1 and 2 in free space, examined at point $\vec{x}_1$ at time t_1 and $\vec{x}_2$ at time t_2. Each time label is related to a common, standard time t by $t_1 = t + \Delta t_1$, $t_2 = t + \Delta t_2$. If the wave functions multiply, the time dependence is

$$e^{-\imath\omega_1 t_1}e^{-\imath\omega_2 t_2} = e^{-\imath(\omega_1+\omega_2)t}e^{-\imath\omega_1\Delta t_1}e^{-\imath\omega_2\Delta t_2}.$$

The frequencies conjugate to t *add*. We have just discovered the *law of addition of energies* for basic non-interacting systems. The overall energy as frequencies *is the sum* of the subsystem frequencies.

Relate each position label to a common, standard position $\vec{x}$ by $\vec{x}_1 = \vec{x}+\Delta\vec{x}_1$, $\vec{x}_2 = \vec{x}+\Delta\vec{x}_2$. If the wave functions multiply, the space dependence is

$$e^{\imath\vec{k}_1\cdot\vec{x}_1}e^{\imath\vec{k}_2\cdot\vec{x}_2} = e^{\imath\left(\vec{k}_1+\vec{k}_2\right)\cdot\vec{x}}e^{\imath\vec{k}_1\cdot\Delta\vec{x}_1}e^{\imath\vec{k}_2\cdot\Delta\vec{x}_2}.$$

The wave numbers conjugate to $\vec{x}$ *add*. We have just discovered the *law of addition of wave numbers* for basic non-interacting systems. The wave number conjugate to $\vec{x}$ *is the sum* of the subsystem wave numbers.

The result is confirming a rule of beginner's physics that *the total energy is the sum of subsystem energies*. That rule has been obtained for non-interacting subsystems, because that's the only case where it is exactly correct. When systems interact there is an *interaction energy*, which may or may not be small, and needs to be accounted for. Since the only case many have seen is the simple addition case, we've used it to make a point, and also warn you. It is premature to talk of many quantum systems, and when we get there wave functions will not simply add. There will be subtle and interesting consequences of composing wave functions by products. We'll get there when we are ready!

5.3.4 FIAQ

Why did the presentation of the phase velocity not also discuss the group velocity, which is more important? Many students are confused by this. The free-space phase velocity formula is $\vec{v}_{phase} = \vec{k}/2\mu_e$. It is the true velocity of harmonic waves in the expansion, mode by mode. The relation $\vec{k} = 2\mu_e\vec{v}_{phase}$ contradicts the by-rote *MSR* rule $\vec{p}_N = m_N\vec{v}_N$, and there's no way to save it.

When a generic blob of waves starts moving, its center moves, it spreads, and it changes shape in many different possible ways. There exists about seven different formulas for 'group velocity' to describe different features. None are correct for everything: all are wrong for something. However, early in quantum history *somebody* wanted a Newtonian-looking formula to come out[4], and popularized the (best-known) group velocity as 'the answer'. If you see a source calculating with a Gaussian wave packet, be very critical when reading it, and you will discover the wave packet does *not* move at the group velocity, despite language claiming or suggesting it.

The unfortunate outcome was that many students memorizing the group velocity could do one homework problem on group velocity while becoming unable to look at wave functions and understand anything else. We're not going to repeat the argument, nor the formula, because nothing in quantum physics uses the group velocity, and dwelling on it just wastes time.

I have seen eigenvalue equations like equation (5.7) described to be physical principles defining what an experiment measures. Why are they not described that way? Because math facts are not physical principles. Math facts are the outcome of your own assumptions. We don't know what nature is, but it is operating on its own regardless of our assumptions.

There is a dogmatic presentation where eigenvalue equations are introduced as free-standing statements about existence. It is deceptive, and won't work with adults who understand eigenvalue equations. The Copenhagen presentation was loaded with unsupported, free-standing statements, under the excuse that any logical system must start somewhere with some definite axioms. The 'eigenvalue postulate' we mentioned is one such thing. If you have never seen an eigenvalue equation, you are told to accept that 'an experiment always measures an eigenvalue and leaves the corresponding eigenfunction'. There do exist special cases where this occurs: for example, if you work hard enough to filter out a very pure plane wave, the eigenvalue relation of equation (5.7) applies. There are two faults with the organizational flow chart putting this early. First, there's no explanation or attempt at explanation: it is a form of bullying by non-explanation. Second, the eigenvalue relation comes from mathematics, so that when it might apply, it's a math fact, and when it does not, physics will not be affected. Finally the statement 'an experiment always measures' is dead on arrival with the word 'always'. Children can be told with authority what nature must necessarily do in experiments, but nature does not obey human authorities. There is a special term some people use for those real-world measurements that don't obey the authorities. The are called *improper* measurements.

References

[1] Franck J and Hertz J 1914 *Verhandl. DPG* **16** 457

[2] Elsasser W 1925 *Naturwiss* **13** 711

[3] Elsasser W 2017 APS Oral History project https://www.aip.org/history-programs/niels-bohr-library/oral-histories/4590

[4] The perpetrator seems to have been de Broglie, who cites Rayleigh and 'vitesse de phase et vitesse de groupe' in section II of his 1924 dissertation.

Chapter 6

Atoms are musical instruments

Figure 6.1. Light on water and soap, by Professor Hamid Kellay of the University of Bordeaux.

6.1 The quantum clues you never knew

Here is the tale of the burning philharmonic. One day a large orchestra was rehearsing for a major recording. Every violin was tuned exactly like the others, because violinists are so fussy. The trombones and oboes and flutes were adjusted to perfect pitch. The timpani were the last ones made by Stradivaridrum. During a break fire broke out. The musicians were safe, having lunch and cocktails at a restaurant across the street[1]. The beautiful instruments of the philharmonic orchestra burned.

And THEN, goes the tale, the forensic examiners recovered the recording of the burning ensemble. Computer analysis of the chaos showed inexplicable, mathematical regularity, and beauty. The record was astonishing. Pure tones sang with sweet harmony from flaming violins. The woodwinds breathed fiery expiration

[1] Vegan barbeque.

doi:10.1088/978-1-6817-4226-7ch6

through their body cavities. The tortured trumpets played a full octave higher than the score. The inexplicable music of a spontaneously combusted orchestra was pirated and distributed world-wide on the internet, outselling the work of artists[2] invited to the American White House.

The parable does not exaggerate. It under-describes the fantastic experience of anyone seeing the pure colors of hot atoms in a flame, and also understanding it. If you have the resources of the top 1/10% you should buy a university laboratory and experiment around with quantum physics.

6.1.1 Atomic spectra

'Atomic spectra' mean the frequencies of light given off by hot atoms. The 'good' experiment needs the atoms to be separated enough so the light comes straight to your detectors, without multiple reflection and re-emission. That's why atoms of a gas, heated in a flame or by electric currents, are preferred. A 'spectroscope' sends light through a slit, giving it a sharp direction, and then through a prism or diffraction grating. The prism affair bends light according to its frequency. *This is*

Figure 6.2. A wealthy 19th Century gentleman discovers the joy of a spectroscope. You can now buy one for $8. The prism is the triangle shape in the middle. This device is comparing two different spectra. The most amazing discovery is found in between the separated spectral lines. One sees *no light*.

[2] 'Gangnam style', 'Montgomery, Flea Market', etc.

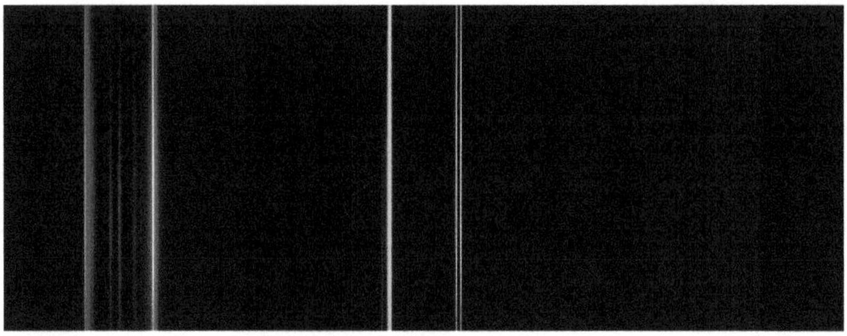

Figure 6.3. A few lines of the spectrum of mercury atoms... dozens of lines exist. The gaps between the lines are the real clue. Photo by Jan Homann.

a quantum measurement: it yields an 'eigenstate of wave number', which like other measurements, works out by the dynamical rules of wave physics without a post-dated postulate claiming it's mysterious.

Looking through a telescope focused on the far side of the prism, one sees a bunch of colored lines, which are copies of the slit (figure 6.3). At first it is confusing why there are so many copies. Then you realize they are images of rays bent according to color. The colors are pure and beautiful, we repeat. The completely non-obvious thing is that between the colored lines there is nothing. Nothing! No light is emitted at *almost all* colors. Each kind of atom only emits light with a certain set of sharp frequencies, and that is simply *bizarre*.

You might be expecting to hear that inherent quantization of atomic energy levels explains this. *Hell no.* That was the non-explanation of the previous millennium. Nothing was ever explained by postulating that the discrete quantization of frequencies that were observed was a *consequence* of postulating that discrete frequencies must be observed.

6.1.2 The unknown history

Well before 1900 the clues of atomic spectra were understood. It is a simple matter of counting the number of physical variables. It is known that *any linear oscillating Hamiltonian system* will have exactly one frequency per oscillating coordinate. For an 'oscillator' you can imagine a ball on a spring. For 13 oscillators you can imagine 13 balls and any number of springs. For your comprehensive exam, you might be asked to solve equations for 13 balls connected by $13 \times 12/2 = 78$ springs, which is as many as can be jammed between pairs, and recover the fact there's still 13 oscillators, and at most 13 different frequencies.

Returning to the atomic spectra, more than 13 lines were observed. There are thousands of lines even for hydrogen, the atom with only one electron. There appears to be infinitely many lines, which turns out to be the truth and the secret of the atom. Well before 1900 this clue was understood. *Waves* are mechanical systems equivalent to *an infinite number of oscillators*. The things of the atom that oscillate are the innumerable bits of the continuous wave, *which can be divided and subdivided ad infinitum.*

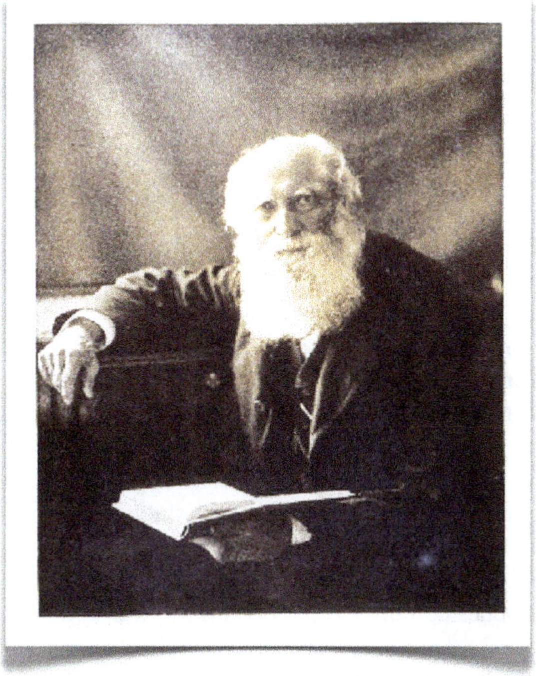

Figure 6.4. George Johnstone Stoney, who discovered the electron and published the solar system model 30 years before it was called the Bohr model.

This is so simple people are not sure they can trust their intuition. If you divide a wave in two, can the left and right halves vibrate independently? Yes, but don't forget they are coupled in the middle. Divide each half in two: each half is independent, to vibrate as it will, while coupled at the boundary. Subdivide the subdivisions forever, and observe there's no limit to the number of shapes, and then the possible number of vibrational frequencies:

$$\psi(x, t) = \sum_{n}^{\infty} c_n \psi_n(x) e^{-i\omega_n t}.$$

Note the sum extends to infinity. The research staff can easily do the math, which to some extent consists of simply making definitions and concentrating on the cases where the definitions apply.

It was not only possible to make the deduction, but the deductions were actually made quite early, in deciding that atoms must be little vibrating blobs of some kind of jello-like jello; since the hydrogen atom is one electron that tells you that one electron must be little vibrating blobs of some kind of jello-like jello. Letters between Stokes, Kelvin, and Lorentz speculated on this in the 19th Century. In 1906 Lord Rayleigh published an 'electron fluid' model of the atom [1], which unfortunately made a poor fit to experimental data. The same year James Jeans commented on Rayleigh's atom [2], writing:

> 'Thus if we regard the atom as made up of point-charges influencing one another according to the usual electrodynamical laws, the frequencies could depend only on the number, masses, and charges of the point-charges and on the aether-constant V. What I wish to point out first is that it is impossible, by combining these quantities in any way, to obtain a quantity of the physical dimensions of a frequency.'

This was true of *Newtonian* physics, which had no intrinsic frequency parameter. Jeans continued:

> 'It seems, then, that we must somehow introduce new quantities—electrons must be regarded as something more complex than point-charges. And when we have once been driven to surrendering the simplicity of the point-charge view of the electron, is there any longer any objection to putting the most obvious interpretation on the line-spectrum, and regarding its frequencies as those of isochronous vibrations about a position of statical equilibrium? The main objection felt against this view is here supposed to lie in being inconsistent with the point-charge view of the electron. The present author suggests that the same objection applies equally to the 'orbital motion' interpretation.'

So Jeans found the flaws in the point charge and the orbital motion interpretation seven years before Bohr kept it in his model.

The date 1906 conflicts with the popular information that Bohr discovered the atom in 1913. *No*, Bohr invented and explained nothing. He simply reverse-engineered the spectrum of hydrogen fit by Rydberg to an unexplained postulate about angular momentum quantization. Rutherford had already popularized the solar system model of the atom with his interpretations of 1911. Rutherford got the model from Nagoaka (1904) and Nicholson (1906). They all acknowledged George Johnston Stoney as the actual inventor of a solar-system atom in the 1880s. Stoney committed to human music so completely he predicted too many atomic frequencies would be harmonic ratios. Arnold Schuster (1851–1934) conducted a thorough survey of spectra which rejected such a simple regularity. That discouraged scientists from working on Stoney's idea. Yet Schuster wrote about 'the inverse problem'[3] in 1881:

> 'We know a great deal more about the forces which produce the vibrations of sound than about those which produce the vibration of light. To find out the different tunes sent out by a vibrating system is a problem which may or may not be solvable in special cases, but it would baffle the most skillful mathematician to solve the inverse problem to find the shape of a bell by means of the sound by which it is capable of sending out. And this is the problem which ultimately spectroscopy hopes to solve in the case of light.'

[3] Marc Kac, a famous mathematician, got other mathematicians excited with the question *'Can you hear the shape of a drum?'* The question comes because each frequency of a well-defined wave system comes with a well-defined shape. Yet the answer to the mathematical question is 'no'.

The *inverse problem* of finding a wave equation whose solutions matched the atomic spectra is precisely what Schrödinger solved in 1926.

6.1.3 The sound of every tune and no particular tune all at once

Figure 6.5 shows the sound spectrum of the singing art-form of the Republic of Tuva in Russia. The spectrum shows a graph of intensity of sound versus frequency. Sharp peaks show high intensity at certain points, with gaps in between. The bars at the top of the figure show evenly spaced harmonics, which the Tuvans work hard to produce by resonances of their vocal system, which sounds rather disturbing to Western ears. Instead of singing one note, the Tuvans are quantum experts who sing 'no particular note, but every note and all notes at once'.

The spectrum of sound from a guitar string ('high-E') is more orderly than Tuvan singing. The repeated bumps are harmonics, which are simple integer multiples of the frequency. A guitar string finds its own frequencies automatically, because it is trapped between fret and a bridge to do nothing else. A trapped electron wave finds its own resonant frequencies where it sustains itself. Generally many vibrational patterns happen simultaneously: that is what waves do.

Figure 6.6 shows the guitar spectrum left to right in the top panel. Below that, we rotated the figure 90° so that each frequency peak looks like a horizontal band, or 'frequency level'. The rotated graph is on the right, with higher frequencies at higher levels. The figure explains almost all there is to know about 'energy levels'. The frequency levels and energy levels are precisely the same thing. We mean the word 'precisely' very literally.

Consider this. If we showed you nothing but energy levels (some levels of mercury are on the left) you might never figure out it is a graphics trick. Rotating a graph of frequency spectrum by 90° makes a graph of frequency levels. Yet millions of students are daily learning that energy levels are intrinsic inexplicable mysteries

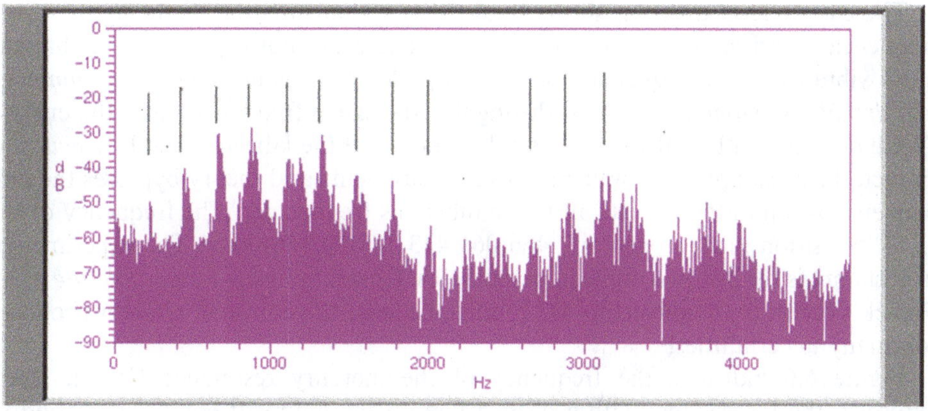

Figure 6.5. The intensity spectrum of a Tuvan singer versus frequency. The vertical lines show harmonics the Tuvans produce. Rather than sing one note, the art form of 'quantum singing' is about singing many notes all at once. Figure by Professor W G Unruh of the Canadian Institute for Physics & Astronomy.

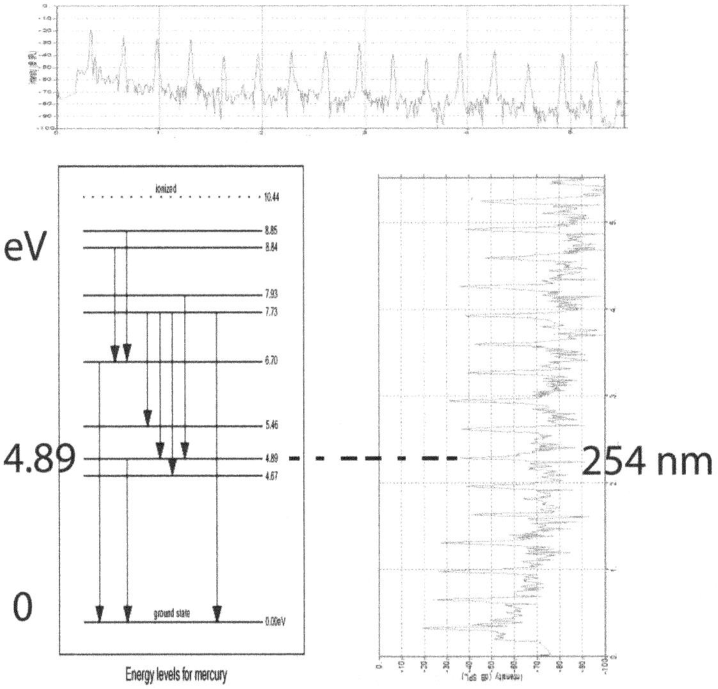

Figure 6.6. The mercury guitar. The top panel shows the spectrum of mercury as a function of frequency. The notation is conventional: intensity on the y axis, versus frequency on the x axis. The bottom panel shows the *frequency levels*, which are simply the frequencies of a plot rotated by 90 degrees.

(Bohr's picture) without ever being told the connection to frequency resonances they can understand.

OK, if we had started off wrong with energy levels, then we would have needed to explain where the frequencies came from. It was ingenious back then: having no physical picture was Bohr's claim to accomplishment.

The energy level diagram shows both the same information as the guitar string frequencies, and also different information. The exact correspondence is that the atom vibrates at the frequencies of the energy levels. It is nothing but *terminology and archaic Newtonian units* that distinguish the sharp frequencies and the energy. (We'll do more with that.) For physical purposes all the labels should be *frequency*. In fact, the most precise atomic physics measurements and theory bypasses the unit conversions, and literally lists all the numbers as frequencies. The frequency of the $1S2S$ transition in hydrogen is 2 466 061 413 187.103(46) kHz. The experimental uncertainty is the last two digits in parentheses. Current physics *cannot tolerate* using Planck's constant to convert to *MKS* units, because the conversion factor produces too many uncertainties!

Figure 6.6 indicates the frequency of the mercury resonance Franck–Hertz explored, the lowest one, with a dashed line on the figure. It could correspond to a high frequency guitar string vibration immediately across the figure, except the scales of frequency are so vastly different. This is all precisely faithful to our present theory, and needs no justification by external authority, which is why you can understand it.

But there is another feature not yet explained. In the obsolete and bad-for-you picture of quantum jumps, the photons supposedly come out by magic with energies which are 'differences of the energy levels'. That is, the vertical arrows on the left side of the figure show the differences of *frequencies* of the atomic levels. That is not how guitar strings work. If a guitar string vibrates at 404 Hz, you hear sound of frequency 404 Hz. This is a clue that the Schrödinger equation will not be the sound equation, but something different. The way it works turns out to be very neat.

6.1.4 The quantum current

When Schrödinger was exploring his equation, he naturally knew it must predict a locally conserved current, so he found the formula right away. It is not entirely clear whether Schrödinger used the relation of symmetry and conservation known as *Noether's theorem* to find the formula. It was generally known at the time and certainly used soon afterwards. The Schrödinger equation has a symmetry under which the wave function ψ and $e^{\imath\theta}\psi$ cannot be distinguished. Superficially this comes from replacing $\psi \to e^{\imath\theta}\psi$ everywhere in the equation, and finding the phase factor just cancels out. Noether's theorem shows that the conservation of electric charge follows from the symmetry. The symmetry is also quite a bit deeper and more intricate than the superficial version. Understanding it more deeply eventually led to the generalizations found in modern strong and electro-weak theories.

Anyway, Schrödinger found the following formula for the current:

$$\vec{j}\,(\vec{x}, t) = -\frac{\imath e_Q}{2\mu_Q}(\psi^*\vec{\nabla}\psi - \psi\vec{\nabla}\psi^*). \qquad (6.1)$$

Here e_Q is the quantum parameter for the overall size of the charge. Notice the current is *zero* if ψ is purely real, *or* purely imaginary, *or* whenever its complex phase is constant ($\psi(x) \sim e^{\imath\theta}|\psi(x)|$). As promised earlier, two parts (real and imaginary) of the quantum wave are needed to make a conserved current.

Observe that the current is smaller when the stiffness parameter μ_Q is larger, with everything else fixed. Then electron waves, which have the smallest μ_Q of charged systems (just as you'd expect) have the most spectacular currents and electromagnetic interactions. (For example, when you look at your thumb, you only see the electrons on the outside. The rest of your thumb is invisible!) Notice that if you know ψ, you can compute the current, but not the other way around. The conserved 'stuff' is not ψ but a measure of the wrinkles in ψ that must move through a bounding volume to get out.

Most of the symbols in equation (6.1) probably do not speak to you right away, but notice this: the current is not proportional to the wave amplitude ψ, but proportional to its square. This means that if you add two waves the current will not add in a simple linear way. One might guess this would have amazing industrial applications, which has been borne out. Another fact: suppose a matter wave vibrates with more than one frequency, which happens to be the typical situation.

For example, some shape $\phi_1(\vec{x})$ vibrates with angular frequency ω_1, and $\phi_2(\vec{x})$ vibrates with angular frequency ω_2. (Such special shapes always exist, as we'll discover in section 7.0.2.) A superposition goes like

$$\psi(\vec{x},t) = c_1\phi_1 e^{-\iota\omega_1 t} + c_2\phi_2 e^{-\iota\omega_2 t}.$$

Let $\vec{j}_1$ be the current calculated with ϕ_1. In computing $\phi_1^*\vec{\nabla}\phi_1$ the time dependent factor $e^{-\iota\omega_1 t} \times e^{\iota\omega_1 t}$ cancels out. This is the same with $\vec{j}_2 \sim \phi_2^*\vec{\nabla}\phi_2$. The total current is

$$\vec{j} = \vec{j}_1 + \vec{j}_2 + const\ Im[\phi_1^*\vec{\nabla}\phi_2 e^{-\iota(\omega_1-\omega_2)t}], \tag{6.2}$$

where *const* involves e_Q/μ_Q and $c_1c_2^*$. Notice the *Im*[] part, which needs a complex wave, and picks out the exact phase relation making a current.

By inspection, the electrical current does not vibrate at a single wave-function frequency, but at the *difference* of two frequencies in the wave function. This explains so much: and it has generally been ignored, or forgotten.

We have seen the equation for the current: by the Schrödinger equation, it is locally conserved with the charge density, which has the formula

$$\rho\vec{x} = e_Q\psi^*(\vec{x})\psi(\vec{x}).$$

The symbols $\psi^*\psi$ are very famous and often introduced in a sort of oracular presentation as 'the probability to find the quantum particle'. Besides the point *parcle* objection, let's observe that the claim omits the information about local conservation. In fact, local conservation contradicts the common idea that the point *parcle* jumps around randomly and might appear anywhere. Since that idea was wrong, it's a good thing to contradict!

6.1.5 Light has the beat

As we said, the wave equation for light has no preferred frequency. Light adapts itself to matter and is emitted with the frequency of the electromagnetic current. In one simple step the current of the Schrödinger equation explained the spectral emission frequencies of light. A hot disturbed atom is thrown into a mixture of internal vibrational frequencies. It is just what you would expect for a little blob of quantum jello crashing about. The electric current vibrates at the differences, or *beat frequencies*, of the electron wave frequencies. A vibrating atom is like a radio antenna, emitting light at difference frequencies until the wave runs out of differences and reaches a pure frequency. The beats stop if the atom goes to its lowest frequency condition, called the ground state. The beats also tend to be weak if the atom transiently vibrates near any single frequency, the natural frequency level of the atom. The phenomenon of light 'getting the beat' effortlessly explains why atoms cascade down in energy over time while emitting the sharp spectral frequencies observed.

Yet this beautiful discovery was largely spoiled by the previous non-discoveries of the Bohr model. The model *did not* discover that atomic spectra are the differences of frequencies: that was already in Rydberg's empirical fit, and already generalized

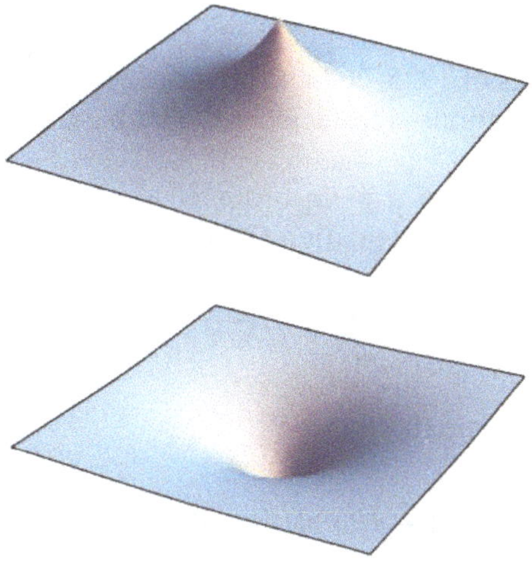

Figure 6.7. Two snapshots of a bouncing hydrogen atom. The atom is three-dimensional and spherically symmetric. The artist making the picture transcribed the amplitude to the 'height' of a wave which is easier to render.

to all spectra, by the so-called Rydberg–Ritz combination principle of 1908. The Bohr model actually assumed more inputs than its outputs. Assume special 'orbits' exist, assume instantaneous jumps, assume there are photons with $E = h\nu$, and assume conservation of energy. Notice there is no mechanism, no element of time evolution, and no mention that light needs to obey its own wave equations that are driven by the electric current. The assumptions are not intended to stand on their own: they were formed by working backwards from data, and *decreasing* the information given by data. That is why *Wikipedia* will report that Bohr predicted the hydrogen atom spectrum.

In this century you need to know it is a serious conceptual error to use 'conservation of energy' as a fundamental principle. You can salute it all you want: conservation laws are *outputs* of equations of motion, which have more information. If you have an equation of motion you can calculate everything, which will already be consistent with a conservation law, while having full detail, such as the amount of time needed for an atomic transition. The time Δt a transition takes is observable with the relation $\Delta t \Delta \omega \geqslant 1/2$, where $\Delta \omega$ is the observed spread of frequencies in spectral lines. (Line widths had been classified and named *s*, *p*, *d*, *f* for sharp, principal, diffuse, and fine years before the quantum models.) Meanwhile with a conservation law you might anticipate a numerical relation, which the law itself does not actually explain, and you cannot compute time evolution.

The Schrödinger current is what makes all electrical technology possible. The typical atomic vibrational frequencies of electron waves are so high—10^{15} Hz and above—we could hardly communicate with them. However, the *difference* of vibrational frequencies can extend all the way down to zero, namely DC and 60 Hz AC. The picture of the Bohr model that atoms were immune to frequencies below their resonant spectral transitions was *terribly wrong*. If it were correct, 60 Hz AC power would not exist!

6.2 The Schrödinger equation

Schrödinger's notebooks of 1925–26 have been preserved, and they show that more than trial-one wave equations were explored. The basic issues can be understood very easily.

The equation of motion of a plain ordinary *dimensionless* classical wave with amplitude $\phi(\vec{x},t)$ was shown in equation (3.1), repeated here:

$$\frac{\partial^2\phi}{\partial t^2} = c^2\vec{\nabla}^2\phi. \tag{6.3}$$

The 'generic linear wave equation' is

$$\frac{\partial^2\phi}{\partial t^2} - c^2\vec{\nabla}^2\phi + \omega_*^2\phi - \tilde{U}(\vec{x})\phi(\vec{x}) = 0. \leftarrow \text{generic linear waves.} \tag{6.4}$$

The plain ordinary dimensionless equation is a special case setting the parameters $\omega_* \to 0$ and $\tilde{U}(\vec{x}) \to 0$.

For reasons Newton never suspected, these equations have a *generalized Newtonian* form. On the left side is the *acceleration* of the amplitude. A wave induced in a Newtonian mechanical medium will have every little bit accelerating proportional to the local force on it. The right hand side of the equation is the local force of springiness in bending the wave: that is computed from $\vec{\nabla}^2\phi$. While the Newtonian acceleration formula $a = f/m$ appears, it's not quite right to say this comes from Newton. *Any* dynamical equation is going to have some time derivatives.

The most simple solution comes from $\tilde{U}\to 0$ and $\vec{\nabla}^2\phi_{flat} = 0$. Then

$$\begin{aligned} \frac{\partial^2\phi_{flat}}{\partial t^2} &= -\,\omega_*^2\phi, \\ \phi_{flat} &= c_+e^{-\iota\omega_*t} + c_-e^{\iota\omega_*t}. \end{aligned} \tag{6.5}$$

Here $c_\pm$ are initial conditions. The equation describes a certain harmonic oscillator with frequency ω_*. The Hooke's law force $-\omega_*^2\phi$ shows the parameter ω_*^2 is a 'spring constant' tending to restore a wave to zero amplitude. The function $\tilde{U}(\vec{x})$ is a spatially varying version of a spring constant. One anticipates that waves can potentially be trapped in a localized region of 'soft springiness' where $\tilde{U}(\vec{x}) < 0$. Note that $\tilde{U}$ should not have a constant part, in order not to over-count the ω_*^2 term. Then $\tilde{U}(x) \to 0$ as $|\vec{x}| \to \infty$ relative to the interaction region will describe the approach to free space.

Schrödinger went a long ways with this line of argument, finding results about two years ahead of the field before he published anything. What he published used an approximation from the frequency parameter ω_* being enormously larger than atomic frequencies. When ω_*^2 is so large it swamps the other terms in equation (6.4), its effects should be factored out. Make a change of variables

$$\phi(\vec{x}, t) = e^{-\imath\omega_* t}\psi(\vec{x}, t) \leftarrow \text{defines } \psi;$$
$$\frac{\partial^2\phi}{\partial t^2} = \left(\frac{\partial^2\psi}{\partial t^2} - 2\imath\omega_*\frac{\partial\psi}{\partial t} - \omega_*^2\psi\right)e^{-\imath\omega_* t}.$$

The overall phase $e^{-\imath\omega_* t}$ factors out of the other terms in equation (6.4), and cancels out. The transformation cancels out the ω_*^2 term completely, leaving

$$\frac{\partial^2\psi}{\partial t^2} - 2\imath\omega_*\frac{\partial\psi}{\partial t} - c^2\vec{\nabla}^2\psi + (\omega_*^2 - \omega_*^2)\psi - \tilde{U}(\vec{x})\psi = 0,$$

$$-\imath\frac{\partial\psi}{\partial t} - \frac{c^2}{2\omega_*}\vec{\nabla}^2\psi - U\psi \sim 0, \leftarrow \text{the Schrödinger equation!} \qquad (6.6)$$

$$\text{where} \quad U = \frac{\tilde{U}}{2\omega_*}, \quad \text{and assuming} \quad \left|\omega_*\frac{\partial\psi}{\partial t}\right| \gg \left|\frac{\partial^2\psi}{\partial t^2}\right|. \qquad (6.7)$$

As the flag proclaims, equation (6.6) is the Schrödinger equation. For the electron $\omega_* \to \omega_e \sim 7.8 \times 10^{20}\text{s}^{-1}$ is truly huge on the atomic scale, as assumed. The second line of equation (6.7) restricts its applicability basically to situations where the frequencies of ψ are small compared to ω_e. The low frequency approximation is a rough estimate, or 'highly suspicious step' (hss). Even if those conditions hold, a relatively small term in a differential equation will often tend to become non-negligible after a sufficiently long time.

Amazingly, Schrödinger published the *approximate* equation (6.6) with no mention of its potentially more exact predecessor of equation (6.4). He did this despite the fact that equation (6.4) was consistent with special relativity (assuming a consistent interaction $\tilde{U}$) and special relativity was assumed to be universally true in 1926. Schrödinger actually *butchered* his equation with the highly suspicious step (*hss*), *because he was so sensitive he knew butchery was needed.*

Continuity again. Without proof we claimed equation (6.1) was conserved, with charge density $\rho(\vec{x},t) = e_Q\psi^*\psi(\vec{x},t)$. Take the divergence of equation (6.1), and you will see terms of the form $\psi^*\vec{\nabla}^2\psi$. Take the time derivative of ρ, and there will be terms like $\psi^*\partial\psi/\partial t$. Use the Schrödinger equation (6.6), and show that $\vec{\nabla}\cdot\vec{j} = -\partial\rho/\partial t$, to verify the equation of motion predicts the current is conserved. This is easy, and so satisfying we leave it to you to explore.

6.2.1 The artful mutilation of a theory

The problem demanding mutilation appears with the flat wave, equation (6.5). There are two solutions, which mean two distinct ways to cancel out the high frequency oscillations, by defining $\phi \to e^{\mp\imath\omega_* t}\psi_\pm$. The equation for the alternative sign simply replaces $\omega_* \to -\omega_*$:

$$\imath\frac{\partial\psi_+}{\partial t} = -\frac{c^2}{2\omega_*}\vec{\nabla}^2\psi_+ + U\psi_+ \leftarrow \text{the Schrödinger equation,}$$

$$\imath\frac{\partial\psi_-}{\partial t} = \frac{c^2}{2\omega_*}\vec{\nabla}^2\psi_- - U\psi_- \leftarrow \text{the anti–Schrödinger equation.}$$

Equation (6.4) has two inequivalent low frequency approximations, suggesting it inadvertently represents two distinctly different quantum systems. The doubling was already visible in equation (6.5) provided ϕ was complex. It is the solution for *two separate* real harmonic oscillators. (The solution for one real oscillator in complex notation would use the real part of the equation.) We've seen in section 3.1.2 that a conserved current needs a complex wave. Therefore, both ϕ and ψ must be complex[4].

Whenever the equation for ψ_+ is solved, there will be a solution for ψ_- replacing time $t \to -t$. (This is *not* time-reversal symmetry, which refers to a property of ψ_+ on its own.) Schrödinger went so far as to solve the vibrational frequencies of the hydrogen atom for the ± cases, with the predictable result that they occurred in opposite-sign pairs. The physical meaning of the sign of the frequency (section 5.3.1) convinced Schrödinger he should choose the equation for ψ_+ and its solutions matching experiments.

Once he abandoned equation (6.4) as a starting point, Schrödinger needed a 'derivation' of his equation, and especially a derivation that might be acceptable to the mindset of the *OQT* he intended to overcome. Schrödinger then reviewed and modified W R Hamilton's method relating wave optics to classical mechanics, also called Hamilton–Jacobi theory. The derivation should be a side issue because anything truly new in physics is never derived, but guessed. However, pretending there was a derivation became a major issue for the *repackaging* of Schrödinger's theory by those so heavily invested in the *OQT*. That's why the dominant presentation differs from ours, and diligently seeks to make Newtonian physics appear to have relevance and predictive power.

Antimatter! About a year after Schrödinger published, Oskar Klein and Walter Gordon published the *Klein–Gordon equation*, the generic equation (6.4). To their chagrin they rediscovered the same frequency doubling and difficulty of interpretation Schrödinger had found, and could not make physical sense of their own equation. The objective of the 1928 *Dirac equation* was to get rid of those problems. Yet Dirac *increased* the number of frequency-doubled complex waves, apparently making things worse. Dirac fought for his equation to survive, making several mistakes of interpretation which were criticized by Weyl, Oppenheimer, and others. By 1931 Dirac had no option but to propose a new species of matter must exist. Soon afterwards [3] in 1932 Carl Anderson observed tracks in a cloud chamber consistent with antimatter, writing:

'Up to the present a positive electron has always been found with an associated mass 1850 times that associated with the negative electron .... The specific-ionization

[4] The overall phase relating ϕ to ψ cancels out in the formula for the current, so if one is complex they both are.

is close to that for an electron of the same curvature, hence indicating a positively-charged particle comparable in mass and magnitude of charge with an electron.'

Looks like protons were called positive electrons in those days! The name 'positron' for the anti-electron was suggested by an editor of a *Physical Review* paper Anderson published a few months later. At least two scientists had observed the positron before Anderson: in 1929 Chung-Yao Chao and Dmitri Skobeltsyn independently noted electron-like signals with the wrong sign. Lacking a theory to explain it, they did not follow up.

After we know anti-matter exists, there comes a question: why don't we return to the generic equation (6.4), which has the advantage of consistency with special relativity, and avoid the butchery of the *hss*? One answer is that we can, and we have shown you how. The *actual* quantum wave is the one from the Klein–Gordon or Dirac equation, which was replaced by defining the Schrödinger wave function $\psi = e^{\imath\omega_* t}\phi$, *as a convenience and approximation.* If you are impressed by the high frequency of atomic electrons, you need to *add the constant frequency* ω_e to get the true frequency of ϕ. It's very nice this completely explains 'rest mass'. It is the frequency parameter of the flat electron wave in free space: a spring constant. Just as the rest mass was an invisible component of energy in non-relativistic physics, so is the frequency difference between ψ and ϕ.

A more traditional answer is that basic quantum mechanics has traditionally been based on the ordering of the Copenhagen presentation. It totally collapses with the Klein–Gordon or Dirac equations. It is no longer possible even to PRETEND that $\psi^*\psi(x)$ is the probability for a 'quantum *parcle*' position. The lofty axioms contradict themselves *right away* rather than later. (They could not have been lofty when they described an approximate, butchered equation!) So the traditional approach suppresses the whole discussion. When relativity is mentioned in that context, it is done using approximations for small effects arranged not to disturb the non-relativistic framework.

That is not a very satisfactory situation, but most students of quantum mechanics don't hear about it until they come to advanced graduate level coursework. It is interesting that the essential issue does not really come from relativity. The essence is that a second order wave equation—$\partial^2\phi/\partial t^2$—is different than a first order one with $\imath\partial\psi/\partial t$. That is where the destruction happens. It is resolved at the fundamental level by enlarging the space of quantum waves to include all possibilities of matter and antimatter. It is not done by adding in some wave functions for antimatter waves, but by including all necessary *products* of matter and antimatter waves. There was a specific warning in section 5.3.3. It must wait to be developed in chapter 10.

6.2.2 What interaction function?

The Schrödinger equation does not specify the interaction function $U\vec{x})$, which defines the particular system, and which needs to come from physics. There is no shortage of sources solving the Schrödinger eigenvalue problem for a collection of standardly named cases. As discussed in section 5.3.2, assigning the *Newtonian potential* $V_N(\vec{x}) \rightarrow U(\vec{x})$ in appropriate units defines the standard naming scheme.

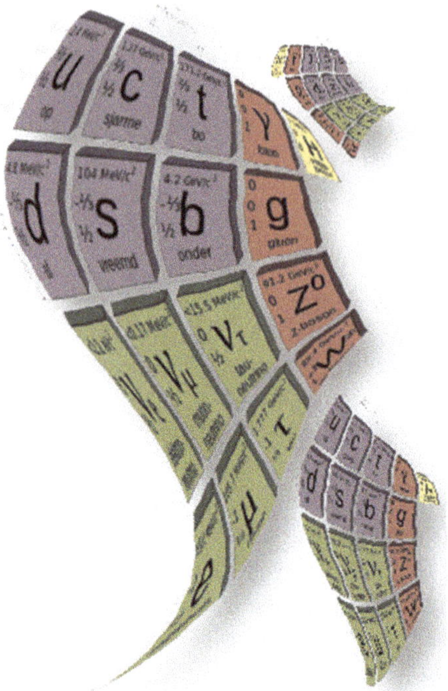

Figure 6.8. Those little boxes pasted up by a committee have no information. At least a committee might have told you those little *par .. cles* were quantum waves of quantum fields.

The Newtonian potential for a one-dimensional harmonic oscillator is $kx^2/2$, so using $U(x) \sim x^2$ is called a 'quantum harmonic oscillator'. Using $U(x) \sim a_1x^2 + a_2x^4$ is called a 'quantum anharmonic oscillator', and so on.

Unfortunately, that resurrects the problems of using Newtonian terms for what a quantum wave does. The quantum harmonic oscillator is not the least bit like a classical Newtonian oscillator, despite receiving the most ingenious efforts of talented enthusiasts wanting to make it appear the same. The older textbooks give the impression one is really accomplishing something by (say) setting up a classical double pendulum, and dealing with equations for the 'quantum double-pendulum'. The idealized, scholastic nature of the exercise is hardly noticed. The positive contribution of solving many Newton-named examples is that one sees many examples.

We have emphasized a different approach. None of this material is particularly 'realistic'. It is all a toy model and prelude to learning how to deal with more physical models. Any given, fixed interaction function can only be an approximation to a dynamical, reacting system with its own life and time evolution. This was already the case in Newtonian physics. When the Moon orbits the Earth, it is treated as moving in a background gravitational potential. Actually the Moon reacts on the Earth with gravitational waves that cause the tides. There is a time-dependent field between the Earth and Moon which is just not *recognized* as gravitational waves.

Since we basically have no idea of the possibilities on the atomic scale, it makes no sense to limit the interaction function. Yet if that makes you nervous, the textbook authorities will solve the dilemma with an authority statement like 'the Schrödinger equation uses the Newtonian potential'.

6.2.3 FIAQ

When you talk about the atomic spectrum, it seems to use classical waves everywhere. Don't you need the fact that photons exist someplace? No, not yet. You are also better off learning quantum mechanics without photons coming too early.

From its inception the notion of the photon as a little point was unworkable, representing a badly designed attempt to reconcile odd kinds of data. Einstein never committed to it. The idea was waiting for clear thinking and good design to be discovered on some different basis. Clear thinking about Nagoaka's atom might have decided the electron current must vibrate at the *difference* of electron frequencies, just as observed. Then, how would that happen?

Already by 1901, Reginald Fessenden had discovered the principle of the *heterodyne receiver*, which combined oscillating signals by *multiplication*. The ingenious facts of multiplication produce sum and difference frequencies. The math identity is

$$\cos(\omega_1 t)\cos(\omega_2 t) = \frac{1}{2}\cos(\omega_1 t + \omega_2 t) + \frac{1}{2}\cos(\omega_1 t - \omega_2 t).$$

To make a current with difference frequencies, Fassenden would have said: 'Just multiply different oscillations'. If *only* difference frequencies and no sums are observed, Fassenden would have said: 'Make phase combinations equivalent to certain complex products'. That is,

$$e^{-\imath\omega_1 t}e^{\imath\omega_2 t} = e^{-\imath(\omega_1-\omega_2)t}.$$

(See the Schrödinger current, equation (6.2).) Yet the bias that the electron was a point-like *parcle* was so strong no one seems to have considered any other formula for the *current*, regardless of the data.

None of this comes from photons nor needs photons. So how do photons fit in to quantum mechanics? Photons are particular excited states of the vacuum. Planck's analysis of the black body spectrum found that for each electromagnetic field wave number $\vec{k}$ there existed a whole number of mode frequencies $\omega_n = nc|\vec{k}|$. When quantum electrodynamics was developed it was found that its nth excited state has the same label n. The main difference with the nth excited state of an atom is that the photon wave states are entirely delocalized. Feynman described a mental block concerning the operator notation describing this in approximate calculations. The textbook photon state is so utterly delocalized it is *everywhere*, which by causality needs infinite time to form. The resolution is that the electromagnetic field finds the best resonance possible, which might be a teensy bit mismatched from the ideal.

That is called a 'virtual particle', which is a double-misnomer. First the thing actually exists, and not only in virtue. Second, it's not a *parcle*.

References

[1] Stutt J 1906 *Phil. Mag.* **11** 117–23
[2] Jeans J 1906 *Phil. Mag.* **11** 604
[3] Anderson C D *Science* **76** 238

How to Understand Quantum Mechanics

John P Ralston

Chapter 7

Waves with known solutions

Figure 7.1. The Schrödinger theory of hydrogen IS small enough to write on the edges of a postage stamp.

Quantum mechanics is easier than you might think, because *the properties of the solution to quantum dynamics are generally known in advance.* Dynamics describes the time evolution of physical objects. The *decision* to put attention on a linear wave equation interacting with a given interaction function has consequences. The general

doi:10.1088/978-1-6817-4226-7ch7

solutions are linear combinations of products of harmonically varying time dependence multiplied by special 'shape' functions. It may take work to find the functions, but mathematicians and physicists often call that a 'trivial' problem. The word trivial means a solution *exists* with a definite form, which happens to be the most useful information possible. In quantum mechanics the general facts are *much more useful* than the details of solutions, even when you are concerned with details! Schrödinger made a very good decision choosing to explore a 'trivial framework' for quantum dynamics. We probably cannot get much information from systems that are not trivial.

7.0.1 A general ansatz

The free quantum wave shows the general pattern of quantum wave solutions. First there was the *trick of no information* and equation (5.4). A general expression about *any* initial wave shape is written down using some general expansion, which symbolically includes initial conditions. The general expression is called an *ansatz*, which means an empty mathematical expression that waits to be completed. The ansatz is a mathematical parameterization not based on any underlying theory or principle. Completion comes when the empty expression is required to comply with the time dependence of the equation of motion.

A time $t = 0$ consider

$$\begin{aligned} \psi(\vec{x}, t=0) &= c_1\phi_1(\vec{x}) + c_2\phi_2(\vec{x}) + \cdots c_n\phi_n(\vec{x}) \\ &= \sum_n^{all} c_n\phi_n(\vec{x}) \leftarrow \text{parameterization, not prediction} \end{aligned} \tag{7.1}$$

Here $\phi_n(\vec{x})$ are a number of 'basis' or 'shape' functions, waiting to be specified. The expression reminds us of a Fourier expansion that uses cosine-shaped waves, which is a special case. The symbols c_n stand for initial conditions for the wave. They are called 'expansion coefficients', while the whole expression is called 'an expansion of a function in the basis ϕ_n'. As the note says, it is a parameterization, not a prediction.

The notion of a basis comes from linear algebra, extended to describing functions. Every basis function must be linearly independent, so none of the coefficients double-count. Many interrelated statements about a basis center on a property of *completeness*, whereby the weighted sums are sufficient to parameterize any functions wanted. To be formally complete a set must be infinite, because infinite variety can exist for functions: there also exist criteria for expressing completeness of particular classes of functions named after dead mathematicians (Banach space, Hilbert space, etc). You are not always obliged to deal with infinite sets, and narrow distinctions specifying 'allowed functions' tend to fail in physics where problems will define their own 'spaces' circularly. That is why the $\sum_n^{all}$ above runs up to *all* the functions wanted, whatever that is. We often discuss cases where just a few terms suffice.

Linear algebra gives a geometric interpretation to functions as vectors, which is by far the most powerful thing to know after calculus… or even better than calculus. It's very annoying that elementary coursework usually does not include the information, which really is quite basic. (Mathematicians are often unaware of what physicists need to know, and distracted by exceptions, and their own ideas of rigor.) Here are three basis vectors written as lists you can also graph as functions, plus one function-list which is a linear combination

$$e_1 = (0, 0, 1, 0, 0, 0, 0\ldots)$$
$$e_2 = (0, 0, 0, 1, 0, 0, 0\ldots)$$
$$e_3 = (0, 0, 0, 0, 1, 0, 0\ldots)$$
$$f_x = -1e_1 + 2e_2 + 7e_3 = (0, 0, -1, 2, 7, 0, 0\ldots)$$

Linear combinations of the e_n basis elements can be rearranged and combined to make smooth basis functions ϕ_n we generally have in mind for smooth, continuous waves. There is not much more to *quantum mechanics* than the Schrödinger equation, applied linear algebra, and a few probability rules, developed *after* you understand the basics.

Function-lists There is nothing more powerful than considering functions to be pre-evaluated lists with indices $f(x) \to f_x$. This is so powerful we recommend you abandon the alternative definition that functions are maps from an input to an output. We are not concerned with evaluating arithmetic. Instead imagine a computer evaluates every function for you the instant it is defined, and stores the result. An example would be

$$f(x) = x,\ x > 0; \to f_x = (1, 2, 3, 4, 5, \ldots)\Delta x,$$

where Δx is a small interval along the x axis. Then $g(x) = 2f(x) \to 2f_x$ makes a list where every element is doubled. By this act of notation, all the theorems of linear algebra can be transferred wholesale to functions. For the sum of two lists, you add the number in the same slots, with the rule for adding functions, and so on.

The function-list concept is very useful because it more than doubles your insight into the math while cutting the work in half. We will *always assume* you have a hard-wired computer, or the equivalent 'research staff' to pre-evaluate functions. The research staff will competently turn wave shapes into lists, and back again, leaving the research boss no work beyond thinking about the lists. The ansatz of equation (7.1) that tends to intimidate simply says: 'I've got a list, which is a sum of some other lists', and that is so easy.

The *norm-squared* of a list-vector $|f|^2$ is defined by the Pythagorean theorem:

$$|f|^2 = |f_1|^2 + |f_2|^2 + \cdots |f_n|^2.$$

Applying this to a function-list gives

$$|f|^2 = |f_1|^2 + |f_2|^2 + \cdots |f_n|^2(\Delta x) \to \int_{-\infty}^{\infty} dx\, |f_x|^2 = \int_{-\infty}^{\infty} dx\, |f_x|^2.$$

It is possible to be fussy over the transition from the discrete sum with $\Delta x \to 0$ to the integral over the whole line. All fussy issues are resolved by knowing the right hand

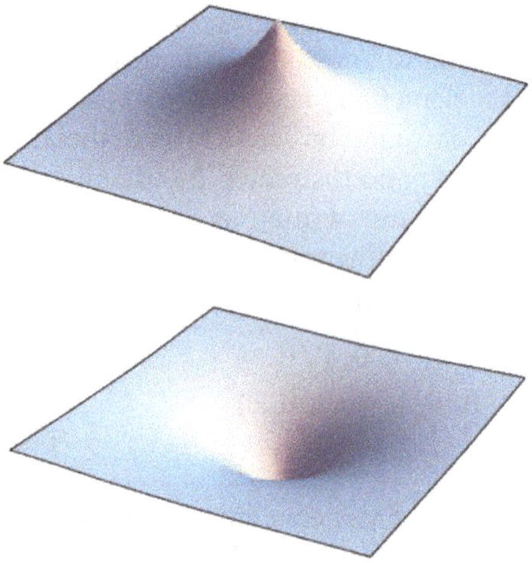

Figure 7.2. Two snapshots of a bouncing hydrogen atom, repeated from figure 6.7. The artist making the picture transcribed the amplitude to the 'height' of a wave, which is easier to render. The figure suggests an expansion consisting of just one frequency eigenstate.

side will be our *agreed definition* of the norm-squared of the function, which will work because it has all the properties of positivity, and so on needed. With the norm we can talk about 'larger and smaller' functions more competently than comparing them at single points. In fact, the norm (or *normalization*) is just one number, which we can separate and consider independent of the other features defining the function 'shape'. Think about this accomplishment: the elusive concept of 'shape' has already been sorted into 'size' from the norm, with the rest being the shape for a given size.

More will come. We want your focus on the shapes of quantum waves and how we describe them with lists, and in particular, lists for shapes that change over time.

7.0.2 The silver bullet: one rule to solve them all

Mathematics cannot tell you which expansion to choose. We come to the 'Silver Bullet',[1] which describes the general solution to quantum dynamics. For any given Schrödinger equation, *there always exist* some special basis functions ϕ_n where the time dependence is

$$\begin{aligned}\psi(\vec{x},t) &= c_1\phi_1(\vec{x})e^{-i\omega_1 t} + c_2\phi_2(\vec{x})e^{-i\omega_2 t} + \ldots c_n\phi_n(\vec{x})e^{-i\omega_n t}, \\ &= \sum_n c_n\phi_n(\vec{x})e^{-\omega_n t} \leftarrow \textit{Silver Bullet}.\end{aligned} \tag{7.2}$$

To repeat, using the ansatz of equation (7.1) with the Schrödinger equation, *one can show there exist* special $\phi_n(x)$ that time-evolve harmonically, by $\phi_n(x, t = 0) \rightarrow \phi_n(x)e^{-i\omega_n t}$. That is *not* a feature of an arbitrary basis expansion, and needs both special frequencies ω_n and (repeat) special ϕ_n, which will be

[1] We invented the name. The Lone Ranger was the star of TV westerns who never missed, because he shot silver bullets.

determined for each system. The special $\phi_n(x)$ are called 'eigenfunctions' or 'frequency eigenfunctions'. You can also interpret the relation as updating each coefficient $c_n \to c_n e^{-\iota\omega_n t}$. For simplicity the Silver Bullet and the general ansatz (7.1) have used the same symbol ϕ_n. Then at $t = 0$ the oscillating factor $e^{-\iota\omega_n 0} \to 1$, and the initial conditions are compatible.

Consider figure 7.2. The ground state shape of the hydrogen atom is described by $\phi_1(r) = exp(-r/a_0)$, where $a_0 \sim 0.5$ Å. The shape is spherically symmetric, namely a function only depending on $r = |\vec{x}|$, with a cusp at the origin: the ground state wave is shaped like a 'pimple' in the quantum stuff. The ground state wave vibrates at the terrifically high frequency $\omega_1 \sim -10^{15}\mathrm{s}^{-1}$: So $\phi_1(r, t) = exp(-r/a_0)e^{-\iota\omega_1 t}$ and $Re[\phi_1(r, t)] = exp(-r/a_0)\cos(\omega_1 t)$. Hydrogen has another special shape $\phi_2(r) = r\ exp(-r/a_0)$ that vibrates at a somewhat slower super-high high frequency, and so on. There is much more experimental information in the special shapes than in the frequencies. If you start with an arbitrarily different shape, the vibrations will be mixed up, and appear disorganized: in fact, that is the prediction of the $\sum_n c_n \ldots e^{\iota\omega_n t}$ form.

The Silver Bullet is both a general solution, and an organizational tool. It is not a calculational task waiting for you to carry out, and it is not a puzzle. It is a form of final answer, from which experimental predictions can be found and tested. The Silver Bullet is a symbolic expression, full of information and also ready to manipulate with other expressions. The general fact of the solution captures the qualitative features, and is much more useful than any particular solution.

It is exceptional in physics to know the solution to dynamics in advance. Solving dynamics is usually difficult, and many dynamical systems have no known solutions. In very simple cases, such as a Newtonian pendulum, the time dependence cannot be expressed in elementary functions. The fact solutions are known comes from the equation of motion being *linear*. The Silver Bullet is a generic math fact of linear dynamical systems, and does not come from quantum physics. And to repeat, there's nothing deep in $\iota = \sqrt{-1}$, whereby the amplitude of the wave function oscillates like an AC circuit in nearly the most simple way possible.

The angular frequencies ω_n are real valued (not complex) by a general agreement not to consider systems whose time dependence would explode exponentially or disappear. OK? (This obvious fact has been presented as a pretentious postulate, 'the Hamiltonian or frequency operator must be Hermitian', or self-adjoint $\Omega = \Omega^\dagger$.) The notation $\sum_n$ does not imply that ω_n are a discrete set, and many important systems have a continuous spectrum of frequencies. If and when a system has a discrete set of vibrational frequencies, separated by gaps, its frequencies are *quantized*. Note that this is a definition, and a circular one, not a principle. The number and values of the angular frequencies ω_n, and the shape functions are what distinguishes one system from another. Part of quantum mechanics simply consists of a bus tour visiting the friendly systems which have exact solutions and checking the solutions put before you. This is much different from scary stories that you'd actually need to *derive* solutions from scratch. We think that's unproductive, and we'll show you how to avoid it.

Every textbook problem that has been solved is a solved time-dependent problem. It is an excellent adventure to type formulas from textbooks into a graphics program and use random initial conditions to make entertaining 'quantum movies', which we recommend.

Example. We mentioned the ground state hydrogen atom eigen-shape function $\phi_1(r) = e^{-\iota\omega_1 t}$. Suppose $c_1 \neq 0$ and all other $c_n \to 0$. Then

$$\begin{aligned}\psi(r, t) &= c_1\phi_1(x)e^{-\iota\omega_1 t},\\ &= (Re[c_1] + \iota Im[c_1])e^{-r/a_0}(\cos\omega_1 t - \iota\sin\omega_1 t).\end{aligned} \tag{7.3}$$

Multiplying the factors will give four terms without changing the information. Suppose $c_1 = 1/\sqrt{2}$ and $c_2 = \iota/\sqrt{3}$. Then

$$\psi(r, t) = \frac{1}{\sqrt{2}}e^{-r/a_0}e^{-\iota\omega_1 t} + \iota\frac{r}{\sqrt{3}}e^{-r/a_0}e^{-\iota\omega_2 t}.$$

Choose $\omega_1 = -1/2$, and $\omega_2 = -1/8$. A movie of the real part of $\psi(r, t)$ will bounce and slosh around like a little quantum wave. If you are able, you *must* make a movie.

7.1 The Schrödinger equation

The Silver Bullet *formally* ends any questioning along the lines of: 'Can you solve quantum mechanics?' We have the solution! The complete solution is equivalent to the Schrödinger equation of motion:

$$\iota\frac{\partial\psi}{\partial t} = \Omega\psi. \tag{7.4}$$

Here Ω is a 'linear operator' with dimensions of frequency that define the particular system equations on the right hand side. Linear operations include multiplication by functions, and also differentiation by x, y, or z any number of times, or any other *linear* operations. Specifying Ω is a *linear operator* means Ω does not depend on ψ, and

$$\Omega(\alpha\psi_a + \beta\psi_b) = \alpha\Omega\psi_a + \beta\Omega\psi_b.$$

When function-lists are considered vectors, every operator has a corresponding *matrix representation*, where the matrix multiplication of a vector gives a new vector.

In section 1.1 we mentioned the simplest linear operation of multiplication by a constant, $\Omega \to \omega_0$. It is *not* a typical example, and useful to explore why. The corresponding Schrödinger equation is

$$\iota\frac{\partial\psi}{\partial t} = \omega_0\psi,$$

which has a solution

$$\psi = c_1 e^{-\iota\omega_0 t}, \tag{7.5}$$

reproducing the Silver Bullet. Note c_1 is not determined by the solution: it is an initial condition, as promised.

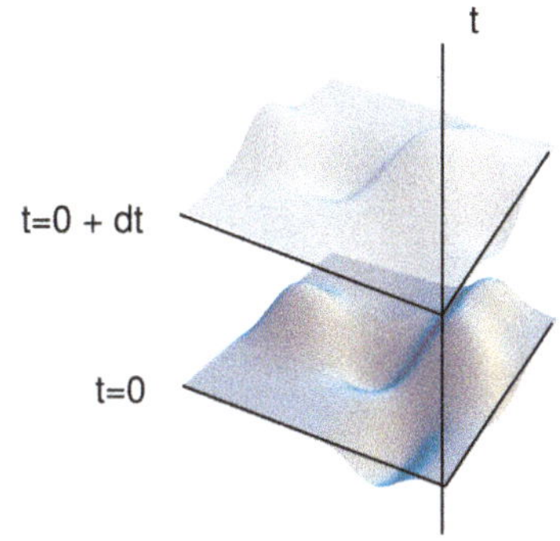

Figure 7.3. How the solutions to PDEs work. An arbitrary, specified wave form at one moment moves to another wave form as time progresses. Every slice of time creates the initial conditions for a new future. Repeated from figure 1.7: the repetition is needed!.

The solution written above is **too easy and incomplete**: it's the kind of mistake inspired by a first order ordinary differential equation (ODE), which always has a single solution. *Waves will never have a single solution.* Waves are described by partial differential equations (PDEs), which are quite unlike ODEs, and have infinitely many solutions. PDEs are so different that one should never show a unique solution, to avoid the impression any exist. (That's why 'cosine solutions' are so treacherous.)[2] The rest of the solutions to equation (7.5) are $\psi(x, t) = c(x)e^{-\iota\omega_1 t}$, where $c(x)$ is any function you choose as initial conditions. Right? equation (7.5) has a sneaky feature that all possible frequencies are the same. Because of that stupid and exceptional property, the generic meaning of the Silver Bullet ($\mathcal{SB}$) to have *special* shapes matched to each frequency is subverted, while the $\mathcal{SB}$ technically still applies.

The details of the Ω operator are specified when the system is specified. You will find disagreements about this: someone will say equation (7.5) is 'not allowed', by setting up rules early to control the narrative. There is a general answer for authority figures who tell you what is 'allowed'. You say: 'Who are you to know what nature will allow!' The real genius of the Schrödinger equation is leaving the frequency operator Ω and even the dimension of the system (number of basis elements) unspecified, until a specific system is confronted. It really should be called the *Schrödinger framework* for quantum dynamics.

The $\mathcal{SB}$ comes from a very simple observation. At time $t = 0$, make an arbitrary expansion $\psi_{any}(x) = \sum_k c_{any-k}\varphi_k(x)$ general enough to express *any* function, given enough coefficients c_{any-k}. The choice of basis φ_k is free at this point and NOT the ultimate symbol ϕ_n. Whatever happens in time will be updated by creating functions $c_{any-k} \to c_{any-k}(t)$. Notice these functions have a Fourier time series, or Fourier expansion in time:

[2] If you read material citing 'separation of variables', burn it. It deceives people to accept unique solutions of ODEs, confusing everything.

$$c_k(t) = \sum_n c_{k,n} e^{-\iota\omega_n t}.$$

To reduce clutter the subscript *any* is implied but not written. Label n keeps track of the frequencies actually present. Swap labels to sum k first and the frequencies last:

$$\begin{aligned} \psi(x, t) &= \sum_n c_n \phi_n(x) e^{-\iota\omega_n t}, \\ \text{where} \quad c_n \phi_n(x) &= \sum_k c_{k,n} \varphi_k(x). \end{aligned} \tag{7.6}$$

equation (7.6) shows that the $\mathcal{SB}$ 'exists' for $\phi_n(x)$, which will be certain special combinations of $\varphi_k(x)$. (We hope you've been distinguishing symbol $\phi \neq \varphi$.)

To find the special shapes, the combination of the Schrödinger equation and $\mathcal{SB}$ is

$$\begin{aligned} \iota\frac{\partial\psi}{\partial t} &= \Omega\psi. \\ \iota\frac{\partial}{\partial t}\sum_k c_k \phi_k e^{-\iota\omega_k t} &\rightarrow \sum_k c_k \omega_k \phi_k e^{-\iota\omega_k t}; \\ \Omega\psi &\rightarrow \sum_k c_k \Omega \phi_k e^{-\iota\omega_k t}; \\ \text{requiring term by term} \quad \Omega\phi_n &= \omega_n \phi_n \quad \leftarrow \text{eigenvalue equation.} \end{aligned} \tag{7.7}$$

The last line is not only 'a solution', it is true for *all* solutions with distinct ω_n, by a fact of linear algebra and the independence of ϕ_n. It also must be true for each label n because the initial conditions c_n are independent. Notice the proof assumes the Ω operator does not itself depend on time, namely it has no time-dependent parameters. This is generally assumed: otherwise, there are few or no useful theorems[3] for system with $\Omega = \Omega(t)$.

The formula above is actually the *theorem of normal modes*, from (generalized) classical mechanics. The theorem says that if you have a linear system, the most general solution is a weighted superposition of special solutions that are eigenfunctions of a linear operator. It's a pity that basic education focused on *par..cle* notions does not recognize *generalized* classical dynamics and the theorem. Schrödinger certainly knew what he was doing when he entitled his 1926 discovery paper [1] 'Quantization as an eigenvalue problem' (*Quantiserung als eigenwertproblem*).

Most information about the system, encoded in Ω, now appears in the particular eigenstates, namely ϕ_n. If ϕ_n is a solution, then $\lambda\phi_n$ is a solution, for a complex constant λ:

$$\Omega\lambda\phi_n = \omega_n \lambda\phi_n.$$

[3] When $\Omega(t)$ is periodic, methods called Floquet theory can be useful.

One says that 'the normalization of eigenstates is not determined by the equations'. To write down a solution one simply *chooses* a normalization (overall scale) of ϕ_n by some agreement. Linear algebra suggests a very clever rule for the normalization, which is always used, but it is still an arbitrary convention. With the convention determined, the initial condition coefficients c_n acquire a corresponding meaning for expansions $c_n\phi_n$.

The minor point of fixing normalizations is sometimes misunderstood to be a special issue of 'quantum mechanics'. The particular error says: 'The norm of any quantum state is 1', on the authority of an early postulate. In the worst presentations, every function encountered will be instantly stopped by the normalization police, and forced to be normalized with integration exercises that demand significant computational effort. Also, if you look up 'eigenfunctions of quantum operators', you will usually find rather complicated expressions, where much of the complication comes from nothing but a normalization convention.

The upshot is that *understanding the origin and meaning* of normalization factors will make learning very much easier. Since eigenfunctions have any normalization you choose, many expressions are best with the simplest normalization possible. With a little judgment one can often postpone normalizations, or efficiently deal with normalizations only at the moment they are needed. Be aware that unthinking followers of tradition will think this efficiency is a *mistake!*.

Untypical example. Computing a derivative is a linear operation. Then d/dx and $\imath d/dx$ are linear operators. For another untypical example consider $\Omega = \imath v d/dx$ in one dimension, where v is a constant with units of length/time. The Schrödinger eigenvalue equation is then

$$\Omega\psi = \imath v\frac{\partial}{\partial x}\psi = \imath\frac{\partial}{\partial t}\psi;$$

$$\imath v\frac{\partial\phi_n(x)}{\partial x} = \omega_n\phi_n(x).$$

The derivative of $e^{\imath k(n)x}$ is proportional to the same function for any $k(n)$: that solves the eigenvalue equation

$$\imath v\frac{\partial e^{\imath k(n)x}}{\partial x} = vk(n)e^{\imath k(n)x} \to \omega_n e^{\imath k(n)x}, \quad \text{hence} \quad \omega_n = vk(n).$$

The index notation needs adjustment: this is very common. Replace $n \to k$ as the label, then $\phi_n \to \phi_k = e^{\imath kx}$. The eigenvalue equation is solved for any value of k, so the allowed $\omega_k = vk$ is a continuous set, called a *continuous spectrum*. Putting the $\mathcal{SB}$ together, the general solution is

$$\psi(x, t) = \sum_k c_k e^{\imath kx - \imath vkt} \to \int dk\ c(k)e^{\imath kx - \imath vkt}. \tag{7.8}$$

The notation $\sum_n \to \sum_k \to \int dk$ evolved as we realized that k was a continuous index. If you enjoy computing integrals, you may concoct initial conditions $c(k)$ that produce a formula you can calculate further: it is best to stop, and reflect.

The example is not typical because the eigenstates came out as $e^{\imath kx}$, the curse of the cosine waves. But the cosine eigenfunctions are just building blocks. Notice the exponent of equation (7.8) is $\imath k(x - vt)$. Since $(x - vt)$ appears as a whole, the final answer will be $\psi(x, t) \rightarrow \psi(x - vt)$, namely some function of x moving rigidly at velocity v. Whatever value of v you choose, perhaps $v = 3.7$, every initial condition time-evolves by moving rigidly at that speed, which is not very physical. To make sure you follow, consider the initial conditions $\psi(x) = 1/(3 + x^2)^2$. Then with $v = 3.7$, the time-evolving function will be $\psi(x, t) = 1/(3 + (x - 3.7t)^2)^2$. These steps bypassed the expression of initial conditions using $c(k)$, which is only an *ansatz* waiting for information after all.

The result is not physical nor interesting, which is why you will not see $\Omega = \imath v\partial/\partial x$ embossed on the Eiffel Tower[4] in Paris as the greatest physical model in history. However, the model is simple, and like an old pocket watch, it shows how the parts mesh together.

All solutions work as we have described. One example has the whole pattern. It was the same pattern with the free quantum wave that used $\Omega \sim (\imath\partial/\partial x)^2$, which we solved with equation (5.10). The free quantum frequency operator has waves moving in both left and right directions, with a different formula for ω_k, and the different eigenvalues and eigenstates were the *only* difference.

7.1.1 Expanding in complete orthonormal sets

The *direct* problem of time evolution specifies initial conditions by choosing coefficients c_n in the frequency eigenstate basis. That is rather easy, because the time dependence is equivalent to replacing $c_n \rightarrow c_n e^{-\imath\omega_n t}$. Since you have the c_n, this is very easy.

The *indirect* problem of time evolution specifies $\psi(x, 0)$, leaving you to find the coefficients c_n. This appears to be a very difficult problem, with infinite variations. Suppose your eigenstates ϕ_n are Bessel functions, and your initial wave is shaped like a top-hat. How will you add Bessel functions to make a top-hat?

Here the property of *orthogonality* solves the problem. Two functions ϕ_1, ϕ_2 are orthogonal when their inner product is zero:

$$<\phi_1|\phi_2> = \int dx\ \phi_1^*(x)\phi_2(x) = 0 \leftarrow \text{orthogonal.} \tag{7.9}$$

Most students first learning quantum mechanics have never seen this, and have no interpretation, *and that is a blight on our friends the mathematicians* for withholding the information. Orthogonality is a supreme test of independence: two functions are completely unrelated when orthogonal.

Orthogonal functions are in every possible way exactly analogous to orthogonal vectors in linear algebra, namely they are at right angles, and independent of each other. It is not difficult to find functions that are orthogonal. Every function that is

[4] Gustave Eiffel put the names of the 72 greatest French scientists, engineers, and mathematicians of all history on the tower. They were all *men*, no women allowed.

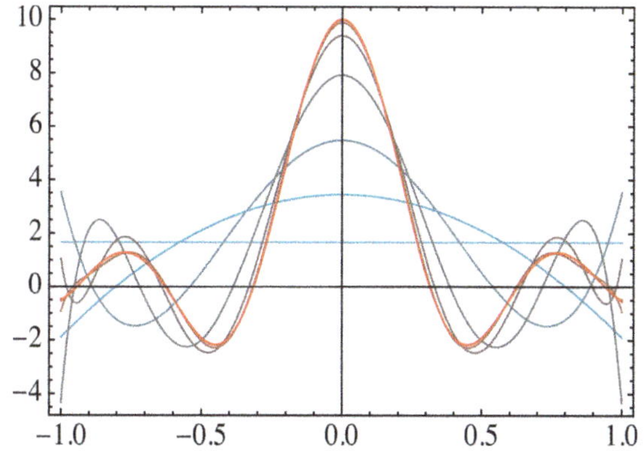

Figure 7.4. Successive approximations reconstructing $\sin(x)/x$ as linear combinations of Legendre polynomials. Their definition is unimportant compared to the fact they are *orthogonal* and *complete.*

even under $x \to -x$ is orthogonal to every function that is odd. However, it takes some work to find two even functions that are orthogonal. *We are not concerned about the labor* of calculating integrals. It is no labor in the 21st Century when a computer can compute a definite integral in a millisecond. Systematic analysis can and has produced dozens of basis sets of mutually orthogonal functions. Many are named after dead mathematicians of the 19th Century, whose lifetime accomplishments were to find orthogonal functions. Move on!

When a set of functions are mutually orthogonal, it is a real accomplishment, written

$$\begin{aligned} &< \phi_m | \phi_n > = const. \delta_{mn} \leftarrow \text{mutually orthogonal} \\ &\text{where} \quad \delta_{mn} = 1,\, m = n; \quad \delta_{mn} = 0,\, m \neq n. \end{aligned} \tag{7.10}$$

This corresponds to a basis of perpendicular 'unit vectors'. Look at this formula and consider $<\phi_m|$ and $|\phi_n>$ to be separate typographical units. When they are mutually orthogonal, all mn combinations connecting one symbol to another at the '|' give *zero*, except when m matches n.

Now *suppose you are so lucky* that your basis expansion, from the beginning, used mutually orthogonal functions ϕ_n. Let $\psi_{given}(x) \to |\psi_{given}>$ be the initial conditions given. The pointy bracket will simplify and automate notation. You want c_m to make the expansion $|\psi_{given}(x)> = \sum_n c_n |\phi_n>$. Due to orthogonality, notice that each particular c_m, (for instance $m = 3$ and c_3), is uniquely filtered or 'projected' from the others by connecting the typographical unit $<\phi_m|$ to the expansion:

$$<\phi_3|\psi_{given}> = \sum_n c_n < \phi_3|\phi_n > = c_3 < \phi_3|\phi_3 >.$$

Inside the sum $<\phi_3|\phi_n>$ gives zero for all cases $n = 1, 2, 4, \ldots\infty$, except $n = 3$. Do you want coefficient c_7? Then connect symbol $<\phi_7|$ to the expansion:

$$<\phi_7|\psi_{given}> = \sum_n c_n < \phi_7|\phi_n> = c_7 < \phi_7|\phi_7>.$$

Projection gives the nth coefficient multiplied by $<\phi_n|\phi_n>$. The basis can always be made to be normalized $<\phi_n|\phi_n> = 1$. (If $|\phi_n>$ is not normalized, then $|\phi_n> / \sqrt{<\phi_n|\phi_n>}$ is normalized.) A normalized mutually orthogonal basis is called an *orthonormal* basis, and obeys $<\phi_n|\phi_m> = \delta_{nm}$. To summarize:

$$\text{Given} \quad |\psi_{given}(x)> = \sum_n c_n|\phi_n>,$$
$$\text{then} \quad c_m = <\phi_m|\psi_{given}>, \leftarrow \text{projection} \tag{7.11}$$

$$\text{assuming} \quad <\phi_n|\phi_m> = \delta_{nm}. \leftarrow \text{orthonormal basis.} \tag{7.12}$$

It is sometimes misunderstood that the inner product $<\phi_m|\psi_{given}>$ will be work to evaluate. The evaluation *has already been done* with the orthonormal basis. It is better to understand that than turn to numerical examples here that *would be work* to evaluate.

All of this assumed the far-fetched luck that one would have an orthonormal basis available. They exist in infinite variety—any orthonormal basis can be rearranged to give any other orthonormal basis—but finding such a basis does not appear to be easy. We happily present you with the *helpful gift of the Hermitian operator*:

THE EIGENSTATES WITH DISTINCT EIGENVALUES OF HERMITIAN OPERATORS ARE ORTHOGONAL.

In particular, since the frequency operator is Hermitian, the eigenstates (with different eigenvalue) of the frequency operator are automatically orthogonal. The Silver Bullet *automatically* uses orthogonal eigenstates; assuming eigenvalues do not coincide. *This is one of the greatest gifts of all time from math to physics.* ***Don't ignore it!***

7.1.2 What eigenvectors mean

Here are facts a person should know from linear algebra. Linear operators make linear transformations of vectors. *Orthogonal* transformations preserve the norm of real-valued lists. *Unitary* transformations preserve the norm $<f|g>$ of complex lists. Symbol $<f|g>$ stands for the inner product. When $f(x) \to f_x$ and $g(x) \to g_x$ are functions-as-lists, the integral creating $<f|g>$ literally makes the inner product. An orthonormal basis is literally a basis of perpendicular vectors, just like a Cartesian coordinate system. Orthogonal and unitary transformations send entire orthonormal basis sets to new orthonormal basis sets.

Symmetric and *Hermitian* transformations are equivalent to scale changes or 'stretches' along special axes. That is the meaning of the eigenvalue equation, $\Omega\phi_1 \rightarrow \omega_1\phi_1$. The eigenvalue is the stretch factor. Most of the information in a vector is in its direction. An eigenvector happens to be oriented so nicely (and matched to a transformation) that the transformation leaves the direction entirely unchanged.[5]

Hermitian transformations have *real* eigenvalues. The stretch property explains geometrically why any eigenvectors with distinct, real eigenvalues are automatically orthogonal. Consider figure 7.5. It illustrates some hypothetical *non-orthogonal* candidates for eigenvectors $|A>$ and $|B>$. Suppose Ω stretches $|A>$ by an eigenvalue $\omega_A = 2$. Then the projection of $|B>$ in the direction of $|A>$ must also be scaled by the same factor. That's inconsistent with $|B>$ being an eigenvector with a different eigenvalue, such as $\omega_B = 3$. One can stretch two vectors by two different factors only when the vectors are completely orthogonal. Hence the theorem: *eigenvectors with distinct eigenvalues of Hermitian operators are orthogonal.*

Using equations, expand A and B in a basis $(\hat{x}, \hat{y})$. Suppose there are distinct eigenvalues $\omega_A = 2$, $\omega_B = 3$. The eigenvalue equations are

$$\Omega A = 2A \rightarrow \Omega\left(A_x\hat{x} + A_y\hat{y}\right) = 2A_x\hat{x} + 2A_y\hat{y};$$

$$\Omega B = 3B \rightarrow \Omega\left(B_x\hat{x} + B_y\hat{y}\right) = 3B_x\hat{x} + 3B_y\hat{y}.$$

The two relations are inconsistent. It is impossible to have $\Omega\hat{x} = 2\hat{x}$ and $\Omega\hat{x} = 3\hat{x}$.

Degeneracy is the term when more than one of the eigenvalues are the same. From the figure it is possible to stretch two different vectors by the same eigen-factor. So the automatic niceness of orthogonality fails. What to do? If you have N independent vectors, you can always make N mutually orthonormal vectors from them. Perhaps this extra work explains the pejorative term 'degenerate', a mathematician's disapproval of eigenvectors not automatically orthogonal. Degeneracy always implies a *symmetry*. The freedom to choose a basis, among a set with no intrinsic preference, is one way to define symmetry.[6] Symmetry is a big and subtle topic, and symmetry does not usually imply degeneracy.

It is very important to understand the geometrical picture of functions as lists, orthogonality, and projection. From high school analytic geometry, the x component of a vector $\vec{W}$ is the dot product $\hat{x} \cdot \vec{W}$, where $\hat{x}$ is one of the *orthonormal* unit vectors of the basis. This should not be a memorized formula, but a self-evident fact. In just the same way, the expansion coefficient $c_m = <\phi_m|\psi>$ should be a self-evident fact. In the remote event your teachers told you to solve a number of simultaneous equations to find expansion coefficients—or if you find yourself going down that road—it's a lot of extra work that takes you 100 years into the past.

---

[5] Defining 'Hermitian' to mean 'all eigenvalues are real' is actually the best definition possible. Eigenvalues do not change under similarity transformations. The self-adjoint test of Hermiticity $H = H^\dagger$ can become unreliable under some similarity transformations.

[6] Given N degenerate frequencies, and making N elements of an orthogonal basis $|e_j>$ by hand out of the eigenvectors, the transformation $|e_j> \rightarrow U|e_j>$ makes an equally new basis, where U is $N \times N$ unitary; and that defines the symmetry.

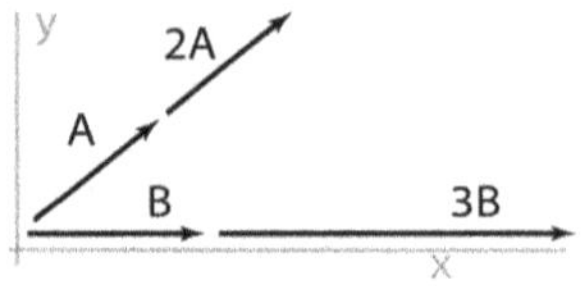

Figure 7.5. It's geometrically impossible for a linear transformation to stretch two eigenvectors by different amounts unless they are orthogonal.

7.1.3 The dogmatic eigenvalue equation

We derived the eigenvalue equation from the Schrödinger equation to make a useful basis of orthogonal functions solving time evolution. Right? The eigenfunctions are an *intelligent* basis, but not more than a basis.

Meanwhile, there is also what we call a 'dogmatic eigenvalue equation', which might give you trouble, written

$$H|\psi> = E|\psi> .$$

When asked for 'the Schrödinger equation' in a comprehensive exam, any physics graduate student writing the dogmatic expression on a blackboard has exactly 30 s to apologize, find the eraser, and replace it by the actual Schrödinger equation.

The mistake happens because pop-culture and many older works have the dogmatic eigenvalue equation bizarrely misidentified as an inexplicable quantum principle. The mistake can be explained in a few steps. First replace $\Omega \to -\vec{\nabla}^2/2\mu_Q + U(\vec{x})$, reducing the generality. Then introduce $\hbar$, in a way to cancel out. Finally rename all the terms:

$$\begin{aligned}
&\iota\frac{\partial\psi}{\partial t} = \Omega\psi,\\
&\to -\frac{\vec{\nabla}^2}{2\mu_Q}\psi + U(\vec{x})\phi_n = \omega_n\phi_n,\\
&-\hbar\frac{\vec{\nabla}^2}{2\mu_Q}\psi + \hbar U(\vec{x})\phi_n = \hbar\omega_n\phi_n,\\
&-\hbar^2\frac{\vec{\nabla}^2}{2m_N}\psi + V(\vec{x})\phi_n = E_n\phi_n, \quad \text{where} \quad E_n \equiv \hbar\omega_n,\\
&\text{often written as} \quad H|\psi> = E|\psi> . \leftarrow \text{never do this.}
\end{aligned} \tag{7.13}$$

Consider the last line 'never do this' first. If you have no background with eigenvalue equations, it does *not* have the information to express a complete fact. It seems to say that whenever you have a wave function ψ, the action of symbol H will always give E times ψ. This, along with the holdovers of the Bohr model, is probably why many have been misled to think every quantum system has a definite energy, symbol E. Omitting the subscript n and the distinction between the wave function and the eigenstate basis functions ϕ_n has been done quite systematically in pop-culture physics. So you should never do it!

In the lines above, defining the symbol $E_n \equiv \hbar\omega_n$ has no information. It comes from multiplying both sides of an equation by $\hbar$, which cancels out however it is written. If you choose to use symbol E_n, it brings the burden of canceling out $\hbar$ every time it is used for physical predictions. Check a few (dozen) textbooks to discover the time evolution factors are written $exp(-\imath E_n t/\hbar)$, which seems to involve $\hbar$, but simply equals $e^{-\imath\omega_n t}$ after $\hbar$ cancels out. We can't stop you from copying $\hbar$ symbols wherever you want them to cancel out, but it's well known that Feynman called it 'a total waste of time'.

You may be tempted to say that $E = \hbar\omega$ is meaningful, because E stands for 'energy', and Einstein proposed the equation to be a principle. Did you notice that we've come to this point without defining the word energy! If you think Newtonian energy would be a definition, NO! It's also not possible to create physical information by making up symbols. Without a consistent definition, the only purpose of introducing a symbol *defined as E identically equal to* $\hbar\omega$ would be to put in an old concept where it has not been needed, *so let's not do that.* When we come to a concept of energy for quantum waves it will stand on its own. It will not come from a symbol circularly defined to make Einstein perpetually correct about things he never considered. (Yet people will believe it anyway.)

Historically equation (7.13) was often disconnected from its derivation, and placed very early as a free-standing principle.[7] The reason was a dogmatic approach claiming that 'measurements will yield the eigenvalue of a Hermitian operator'. It reveals a dramatic difference in ways to *present* quantum mechanics. The claim about measurement does not say what the quantum mechanical system is doing between measurements. That's deliberate, to suppress computable information in favor of a claim one could never know. No time evolution appears in the eigenvalue relation: none existed in the Bohr model. The claim cannot be falsified: when it failed a measurement was 'improper'. They will even misidentify equation (7.13) as THE SCHRÖDINGER EQUATION for stationary states, unaware that 'stationary states' was the term Bohr invented for his model, with the same words implanted in a *presentation* made up after the the new theory had been found.

Let's hope you understand why the dogma is bad for you. Notice you cannot start with the eigenvalue equation, and logically go backwards to understand the mixtures of frequencies that make up smooth and natural time evolution. Right? But if you start with the correct picture of smooth and natural time evolution of superpositions, you see there's a nice theorem that special shapes vibrate with a single frequency, and make good mathematical building blocks for time evolution. *We strongly recommend NO dogma!*

7.1.4 The time evolution operator

The *time evolution operator* formalizes dynamics by defining an operator $\mathcal{U}_t$ such that

$$\psi(\vec{x},t) = \mathcal{U}_t \psi(\vec{x},0).$$

[7] The textbooks by *Liboff* set up eigenvalue equations as quantum principles do.

The operator takes the 'now' state $\psi(\vec{x},0)$ and maps it into 'the future'. This is an elegant idea, with features of seeming to cheat so that $U(t)$ 'solves everything' with no effort. (But notice that the initial conditions $\psi(\vec{x},0)$ are NOT predicted, but supplied as always by YOU.) Once constructed, $\mathcal{U}_t$ will act on ψ with the effect of automatically replacing $c_n \to c_n e^{-\iota\omega_n t}$, *while bypassing all the intermediate steps*, including the explicit determination of frequency eigenstates and eigenvalues. The hook comes when one finds one cannot construct $\mathcal{U}_t$ exactly unless the Schrödinger equation can be solved exactly by some other method such as the Silver Bullet. While it is just another tool, one needs to know about the notion of a time evolution operator, and that it can be useful to develop certain relationships even when it has not been constructed. An example is development of the Heisenberg picture, described in section 9.1.4. We are keeping that material well-separated from this section, because mixing them all together at this point (another confusing tradition) is unnecessary, and just confusing.

7.2 Solved models

There is an abundance of sources presenting a cavalcade of exactly solvable models of quantum systems. Most solutions are competent and one will eventually memorize all of them. We decided to minimize that material here. While practicing math skills with differential equations is absolutely necessary, too much emphasis on differential equations gives a false impression they will be the foundation of understanding. It's not true: linear algebra is far more important, and teaches you that particular solutions to differential equations are not very informative. Another avoidable feature of the traditional approach is computing normalization integrals. We recommend surveying the literature to see how different authors deal with this. Many develop multiple pages of algebra computing normalizations the instant an eigenfunction is obtained. Professional physicists use a trick that every overall normalization is to be postponed, and treated as a garbage can collecting all other pre-factors, just as long as it is possible to get away with it. Since eigenfunctions are eigenfunctions regardless of their normalization, one is being instructed to be incompetent when normalizations get much attention.

Here is a brief review of how a few solved systems work:

The clamped free wave. The frequency eigenvalues of the free quantum wave in one dimension are the continuous set $\omega = k^2/2m$ for arbitrary k. The clamped free wave has the same free space operator $\Omega = -\partial^2/\partial x^2/2\mu_Q$ restricted to the interval $0 < x < L$, with $\psi(x) = 0$ otherwise. This corresponds to 'clamping' the waves at the edges, exactly like the waves in a guitar string are clamped at the bridge and a fret. The *boundary conditions* of clamping quantize the wave numbers. The eigenfunctions and eigenfrequencies are

$$\phi_n = \sin(k_n x) \text{ for } k_n = n\pi/L,\ n = 1, 2, \ldots;\ \omega_n = k_n^2/2\mu \qquad (7.14)$$

The discrete, integer-defined $k_n = n\pi/L$ forces $\phi_n(0) = \phi_n(L) = 0$. No other k values are consistent. Then all superpositions $\psi(x) = \sum_n c_n\phi_n(x)$ obey $\psi(0) = \psi(L) = 0$, as required. Notice that $e^{\iota kx}$ continues to solve the eigenvalue equation for arbitrary k, while almost all k violate the edge conditions. Eigenvalue problems consist of equations *plus* boundary conditions, which together lead to a *vector space* of allowed solutions, and that is the main lesson of the clamped wave.

This is the first example of 'quantization' of frequencies. It does not come from any deep principle. It is a generic fact of trapped waves, whose 'wavelengths' (more generally wave *shapes*) must conform to the trapping conditions. In an atom there is no clamp, but a gentle decrease to zero of the wave outside the trapping zone. Frequency quantization comes from the boundary conditions $\psi \to 0$ as $|x| \to \infty$.

The clamped free wave is often called the 'infinite square well'. The name comes from beginning with the full Schrödinger equation, and choosing the interaction or potential function $U(x) = 0$, $0 < x < L$, $U(x) = \infty$, otherwise. There is also indoctrination language of the form, 'solve the free quantum particle in a one-dimensional box', which turn the almost trivial problem into a puzzle. Moving the clamps to define a different interval $L_1 < x < L_2$ is a typical variation that complicates the expressions and tests whether the quantization concept is understood.

Two- and three-dimensional clamped boxes. The generalization of the one-dimensional clamped wave to mode dimensions is instructive. It is easy to solve the free two-dimensional system with the conditions $\psi(x, y) = 0$ outside the region $0 < x < L_1, 0 < y < L_2$. The eigenfunctions are $\phi_{mn}(x, y) = \sin(k_n x)\sin(k_m y)$, where $k_n L_1 = n\pi$ and $k_m L_2 = m\pi$. The frequency eigenvalues are $\omega_{mn} = k_n^2/2\mu + k_m^2/2\mu$. The general Silver Bullet solution is $\psi(x, y) = \sum_{mn} c_{mn}\phi_{mn}(x, y)e^{-i\omega_{mn}t}$. The appearance of multiple indices was implicit in the Silver Bullet, but not usually appreciated until one works with enough examples on two- or three-dimensional spaces. It is a pleasure to make and see computer-generated movies of quantum 'waves in a box' with perfectly random initial conditions.

Match and patch. For constant U, the eigenfunctions of the operator $\Omega = -\partial^2/\partial x^2/2\mu_Q + U$ are e^{ikx}, while the eigenvalues are the free space eigenvalues plus U. An unlimited number of solvable models can be made by patching together the solutions to piecewise continuous U, making a series of steps or wells. The pedagogical purpose concentrates on *two* conditions that both ψ and its first derivative must be continuous everywhere to solve the second order differential equation. Quantization conditions again come from meeting the boundary conditions. We decided to skip these exercises, because they focus too much on the methods of ordinary differential equations, without contributing much to general skills.

The quadratic potential. The quadratic potential $U = \mu\omega_0^2 x^2/2$ is ubiquitous and probably considered the ultimate prototype of a quantum system. The pedagogical purpose is to display an eigenvalue problem where everything desired comes out sweetly. It is the only quantum system that remotely approaches the classical system, giving it the name 'quantum harmonic oscillator', which it is not. The system is so well known and used that everyone must eventually memorize every possible feature, which (for once) is actually useful, because the model is recycled again and again. A few subtle points are usually overlooked. A slick algebraic method to deduce eigenvalues and eigenstates is actually worse than simply checking solutions, because it gives an impression the method is very powerful, while actually it is over-specialized. Like the clamped wave, the differential equation has solutions for continuous frequencies, while the quantization $\omega_n = (n + 1/2)\omega_0$ comes from the boundary conditions $\psi \to 0$ as $|x| \to \infty$, a fact overlooked by the slick algebraic method. The ground state 'zero point energy' $\omega_0/2$ is meaningless, while many authors thoughtlessly repeat a false claim it 'comes from the uncertainty principle'. It comes from choosing an arbitrary differential equation to solve, while adding any constant to make a different frequency operator simply adds the same constant to the eigenvalues. Since that constant produces an overall phase everywhere, it is completely unobservable.

The continuous spectrum of hydrogen. It's remarkable that the Schrödinger equation for the hydrogenic potential $U = \alpha c/|\vec{x}|$ has analytically solved bound states. They are so famous that many students and professors never think about the states with the continuous spectrum of $\omega > 0$. They correspond to electron waves with frequencies high enough to come in from infinity, scatter off the atom, and exit to infinity. These solutions were found by Arnold Sommerfeld in 1928. While solving difficult differential equations will not necessarily make you a physicist, studying just about everything Sommerfeld did probably will.

7.2.1 Fundamental constants from hydrogen

Imagine we had the shape of the hydrogen wave function from measurements.[8] Then *from experimental data* the ground-state pimple-shape is $\phi_1(r) \sim e^{-r/a_0}$. Experiments give $a_0 = 5.29 \times 10^{-11}$ and frequency $\omega_H = 2.06 \times 10^{16}$ Hz. Given the wave shape, you can ask what differential equation for waves gives the shape as a solution. You could *deduce* the Schrödinger equation.

Wave equations almost always involve $\vec{\nabla}^2$, called the Laplacian operator. When acting on functions of $r = |\vec{x}|$ there is an identity

$$\vec{\nabla}^2 f(r) = \frac{1}{r^2}\frac{\partial}{\partial r}\left(r^2 \frac{\partial}{\partial r}\right) f(r).$$

The identity is written this way to show how it dovetails with fast integration by parts (*FIP*); try it. Then

$$\begin{aligned} \vec{\nabla}^2 \phi_1(r) &= \frac{1}{r^2}\frac{\partial}{\partial r}\left(r^2 \frac{\partial}{\partial r}\right) e^{-r/a_0} e^{-i\omega_H t} = \left(\frac{1}{a_0^2} - \frac{2}{a_0 r}\right)\phi_1; \\ -\frac{a_0}{2}\vec{\nabla}^2 \phi_1 - \frac{1}{r}\phi_1 &= -\frac{1}{2a_0}\phi_1. \end{aligned} \tag{7.15}$$

The left hand side is an operator acting on ϕ_1, and the right hand side is a constant times ϕ_1: we have an eigenvalue equation. Happily, the operator has the Schrödinger

[8] Using measurements to determine a wave function appears to contradict schoolbook lore that measurements must be impossible. Meanwhile the technology has long existed to do the measurements.

form $\Omega = -const\,\vec{\nabla}^2 + \hat{U}$, where a tentative interaction function $\hat{U} = -1/r$ has the wrong units. That is because we started computing $\vec{\nabla}^2\phi_1$ without paying attention to units.

The eigenvalue $1/2a_0$ is an inverse length, not a frequency. Multiply both sides of the equation by αc, where c will convert the units, and α is a dimensionless constant to match the observed frequency:

$$-\alpha c\frac{a_0}{2}\vec{\nabla}^2\phi_1 - \frac{\alpha c}{r}\phi_1 = -\omega_H\phi_1 \text{ s}^{-1};$$

$$\text{where}\quad \frac{\alpha c}{2a_0} = \omega_H; \quad \alpha = \frac{2\omega_H a_0}{c}$$

$$= \frac{2 \times 2.06 \times 10^{17}\text{ s}^{-2}}{2 \times 5.29 \times 10^{-11}\text{ m} \times 3 \times 10^{8}\text{ ms}^{-1}} = 0.0073.$$

The Laplacian of the Schrödinger equation comes with a factor of $-(1/2\mu_e)\vec{\nabla}^2$, or $(-c^2/2\omega_e)\vec{\nabla}^2$. Comparing finds the electron's fundamental frequency parameter:

$$\frac{c^2}{\omega_e} = \alpha c a_0; \quad \omega_e = \frac{c}{\alpha a_0} = \frac{3 \times 10^8\text{ ms}^{-1}}{0.0073 \times 5.2910^{-11}\text{ m}} = 7.8 \times 10^{20}\text{ s}^{-1}. \tag{7.16}$$

Alternatively use $\mu_e = \omega_e/c^2 = 0.87$ s cm^{-2} as the convenient human-scale unit.

7.2.2 Lessons from hydrogen

We have now repeated the analysis done more quickly with table 3.1. The Schrödinger equation for hydrogen is $\Omega_H\psi = \imath\partial\psi/\partial t$, which leads to the eigenvalue equation for the nth state

$$-\frac{1}{2\mu_e}\vec{\nabla}^2\phi_n - \frac{\alpha c}{r}\phi_n = \omega_{nH}\phi_n,$$

$$\text{where}\quad \mu_e = 0.87\,\frac{\text{s}}{\text{cm}^2}; \quad \alpha = 0.0073 \sim \frac{1}{137}.$$

Two fundamental constants μ_e and α come from two pieces of data in the size a_0 and the frequency ω_H. If the fundamental constants are known from other information, the hydrogen parameters are predicted by

$$a_0 = \frac{1}{\alpha\mu_e c}; \qquad \omega_H = \frac{\alpha^2\mu_e c^2}{2}.$$

The interaction function $\alpha c/r$ is rather easy to remember, because it looks very much like the *MKS* Newtonian potential energy $e_N^2/(4\pi\varepsilon_0 r)$. We have enough reviewed the 'cult of belief' that Newtonian physics would predict this: it cannot, so there must be a deeper explanation.

Consider this: electrodynamics itself has no scale. To guess Ω we are looking for a static function $U(\vec{x})$ that must (a) have the dimensions of frequency and (b) be

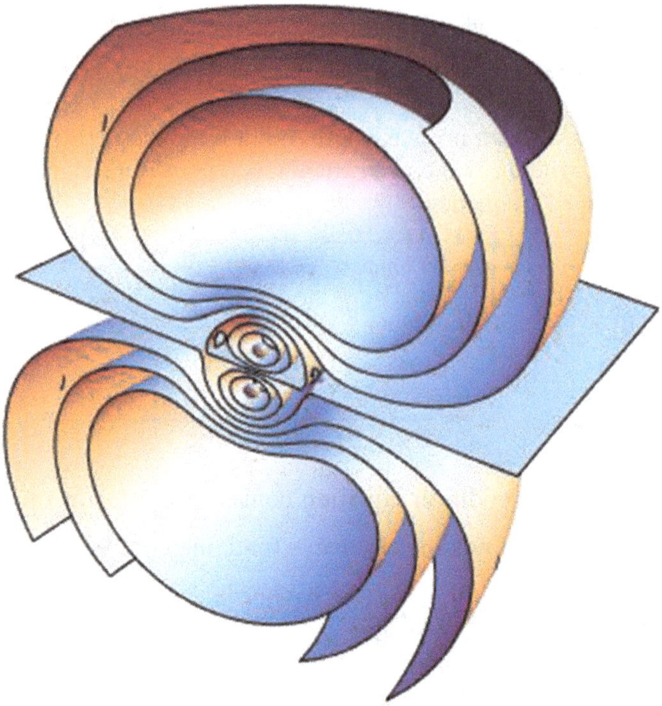

Figure 7.6. Another attempt to look inside a typical hydrogen atom. The artist made artificial layers in the continuum and sliced them in half with a graphics command. The movie is spectacular.

spherically symmetric. From the second item $U = U(|\vec{x}|)$, then $U = c\alpha/|\vec{x}|$ is unique, where α is dimensionless.

By inspection, the constant α sets the strength of the electromagnetic interaction. It turns out to be the *one and only* coupling constant in the theory of electrons with electromagnetic interactions. The whole of quantum electrodynamics (as a theory of quantum electromagnetic and electron fields) has just ω_e and α as parameters. Arnold Sommerfeld called α the *fine structure constant*, from associating it with small atomic spectrum effects he was trying to explain. Sommerfeld was making perturbative series expansions in the pre-quantum theory using *MKS* units. The units were annoying, so Sommerfeld cleverly found a natural dimensionless combination. Since α is dimensionless it refers to no unit system and needs no units: that is how we have found it.

For comparison in the old *MKS* terminology

$$\alpha = \frac{e_N^2}{4\pi\varepsilon_0\hbar c}.$$

This formula is actually faulty if presented as a definition. The formula is only approximate, because the Newtonian parameters e_N and $\hbar$ are too roughly defined and measured to use for precision quantum physics. Modern measurements are arranged so that α stands on its own. Rather than have e_N and $\hbar$ define α, precise quantum measurements of α define e_N and $\hbar$ and Newtonian mass so the *MKS* units cancel out. Right?

Next, the size of the hydrogen atom goes like $1/\mu_e$. Every other atom gets its size from the electron size parameter, so that the whole world of atoms, molecules, and life depends on that number. The muon's intrinsic frequency parameter is 207 times larger, so that if electrons did not exist, the world of muonic atoms, molecules, and muonic life would be 207 times *smaller*, with a few more consequences.[9] Your muonic molecules would want room temperature 207 times higher to make you comfortable: 62 000 K.

7.2.3 More about spherically symmetric systems

We showed directly that *one* eigen-shape $\phi_1 = e^{-r/a_0}$ solved the eigenvalue equation. The shape might have come from experimental measurement, but in fact came from Schrödinger's pen. Fitting two parameters α, ω_e to two data facts does not constitute a test. The tests come when more frequencies and eigenfunctions are found and agree with experiment. That becomes incredibly detailed, going well beyond our purpose here. Here are 'cultural facts' those skilled in the art learn for developing all the solutions:

- Hydrogen is a special case of interaction functions with rotational symmetry, $U = U(r)$. The interaction does not depend on the polar or azimuthal angles θ, ϕ. However, the wave solutions *do* have a wide variety of angular shapes. For any Schrödinger system with rotational symmetry, it can be shown the most general solutions of the Schrödinger equation have the form

$$\psi(r, \theta, \phi; t) = \sum_{n\ell\, m} c_{n\ell m} R_{n\ell}(r) Y_{\ell m}(\theta, \phi) e^{-i\omega_{n\ell} t}. \tag{7.17}$$

  This is the Silver Bullet with detail added. The symbols $Y_{\ell m}(\theta, \phi)$ are famous, known as the spherical harmonics, and have nearly magical (nice) properties. They are developed in the quantum theory of angular momentum. There $\ell(\ell + 1)$ is shown to be the eigenvalue of an operator $\vec{L}^2$ and m the eigenvalue of an operator L_z, identified as the z component of the angular momentum operator $\vec{L}$. This information, for now, is more useful than the demonstration. The index ranges are ℓ = non-negative integer, m = integer, $-\ell \leq m \leq \ell$.
- The problem reduces to solving a one-variable radial eigenvalue equation for functions $R_{n\ell}(r)$:

$$-\frac{1}{2\mu_Q r^2}\frac{\partial}{\partial r}\left(\frac{r^2 \partial R_{n\ell}}{\partial r}\right) + \frac{\ell(\ell + 1)}{2\mu_Q r^2} R_{n\ell} = \omega_{n\ell} R_{n\ell}. \tag{7.18}$$

  This might look difficult, but it is *amazingly simple and easy compared to the scope of the problem posed.* And take heart: we do not expect you to solve

[9] We're neglecting the reduced-frequency correction that involves $\omega_\mu \omega_P/(\omega_\mu + \omega_P)$, which is actually a 10% effect.

difficult eigenvalue problems. The solutions to known problems are actually just *checked.*

- For hydrogen (H) the discrete spectrum eigenvalues $\omega_n = -\alpha^2\mu_Q/2n^2$ do not depend on ℓ. This unusual accident is due to a hidden symmetry, and played an important role in physics history.
- For H the range of ℓ is $0 \leq \ell < n$. For large n one label then hides many eigenstates with the same frequency. The number of possible states has a great effect on the number of transitions and the rates between one another. The Schrödinger equation predicted this spectacularly.
- A pedagogical tradition[10] observes the general solutions for H might be polynomials times e^{-r/a_0}. Somewhat complicated methods deduce the polynomials from recursion relations. The polynomials have long been classified as the associated Laguerre functions. A rather long exercise checks this is true.
- Each nth state has n radial nodes. Larger n states tend to be further from the origin, with effective radii scaling like n. Since ℓ can range up to $n - 1$, a tremendous number of angular variations exist. These atoms spin around like propellors, displaying *angular momentum.*
- Equation (7.17) does not predict the atom must exist in a state of given ℓ. Equation (7.18) solves $R_{n\ell}$ for given ℓ values, but also does not predict an atom must exist in a state of given ℓ. A large number of different ℓ states can vibrate with the same ω_n. These facts show that the hypothesis of intrinsically quantized angular momentum of the Bohr model was dead *wrong*. Since it was wrong, pop-culture will report that quantum mechanics confirmed that angular momentum is inherently quantized.
- Once functions $R_{n\ell}$ are known, they multiply $Y_{\ell m}e^{-\imath\omega_n t}$ to make the general solutions, as already claimed.
- In terms of $\rho = 2nr/a_0$ and dressed up with normalization clutter, the eigenfunctions are

$$\phi_{n\ell m}(r, \vartheta, \varphi) = \sqrt{\left(\frac{2}{na_0}\right)^3 \frac{(n - \ell - 1)!}{2n[(n + \ell)!]}}\, e^{-\rho/2}\rho^\ell L_{n-\ell-1}^{2\ell+1}(\rho) Y_\ell^m(\vartheta, \varphi). \tag{7.19}$$

This has been lifted from sources, and is not well defined until you find the same source's definition of the Laguerre L_n^m, which vary. (If you try it, beware that *Mathematica* tends to get Laguerre indices wrong.)

What did the Schrödinger equation accomplish? Some writers will report that Schrödinger reproduced the results of the Bohr model, which was a great triumph. We disagree, and so should you. Schrödinger reproduced *all the experimental data and more* with a wave equation one could write on a postage stamp. The detail and

[10] Someplace a scholar will object: 'You cannot present the hydrogen atom that way! You must show the complete derivation', perhaps 30 pages. Please! Deriving solutions to differential equations means *checking the solutions claimed* and nothing more.

Figure 7.7. Graphics showing the shapes on the sphere represented by spherical harmonics with $m = l$ for $0 < l < 8$. The plot intensity is proportional to $Re(Y_{lm}(\theta, \phi))$. The spherical harmonics are a terrific *basis* for expanding angular functions, not a discovery of quantum mechanics.

amount of information correctly predicted is just awesome. As far as we can determine, the Bohr model predicted *no results* (except wrong ones) because it was *constructed* from experimental data to circularly reproduce the same data.[11]

Yet despite our enthusiasm, the Schrödinger H-atom is now obsolete, and used only as a first approximation or teaching tool for more than 50 years. After adding some improvements for small effects, the Schrödinger atom reproduced spectra to precision of a few parts per million, but no better. The atom of the *Dirac equation* instantly eclipsed all those accomplishments, and for years was experimentally unchallenged. The framework continued to be an electron wave propagating in a fixed, background interaction function, which unrealistically had no dynamical reaction. In 1947 Willys Lamb and his student R C Retherford conducted a clever experiment to *falsify* the symmetries of the Dirac atom. They observed a subtle and 'slow' microwave beat frequency, now called the Lamb shift, which no modification

[11] A wrong model did not prevent such fine geniuses as Sommerfeld from finding ways to tinker and fit data better and better.

of a fixed, background interaction function could predict. Hans Bethe and many other physicists had been looking for the interaction function to act springy, and show time dependence not found in the static Coulomb type interaction. Bethe produced a quick calculation incorporating effects of a dynamical electromagnetic field and explained the data. The electromagnetic field had always been dynamical: the ingenious step was to isolate the *quantum* dynamical features existing beyond the classical theory.

Since that time hundreds of theorists have calculated many dozen small effects in better and better approximations to test quantum electrodynamics. The term 'Lamb shift' is often applied to all the approximated small terms not found in the Dirac atom. The theory has worked very well. Table 7.1 shows a comparison of experimental *frequencies* with their uncertainties with a state-of-the art calculation.[12] The numerical values to 12 and 14 digit accuracy of the transitions are sufficient to find their definitions with an internet search engine.

7.2.4 Minimal spherical harmonics

We conclude this chapter with some pointers on the spherical harmonics, $Y_{\ell m}(\theta, \phi)$. It is very useful to know what to learn, and what to ignore (for a while) in the vast literature.

Everything important and useful comes from the angular part of the $\vec{\nabla}^2$ eigenvalue equation. The equation is not very useful in the form

$$\vec{\nabla}^2 f = \frac{1}{r^2}\frac{\partial}{\partial r}\left(r^2\frac{\partial f}{\partial r}\right) + \frac{1}{r^2 \sin\theta}\frac{\partial}{\partial\theta}\left(\sin\theta\frac{\partial f}{\partial\theta}\right) + \frac{1}{r^2\sin^2\theta}\frac{\partial^2 f}{\partial\varphi^2}.$$

It is quite useful in the form

$$\begin{aligned}\vec{\nabla}^2 f &= D_r^2 f - \frac{1}{r^2}(\vec{r}\times\vec{\nabla})^2 f,\\ &= D_r^2 f + \frac{1}{r^2}\vec{L}^2 f; \qquad \vec{L} = \vec{r}\times(-\imath\vec{\nabla}),\end{aligned} \tag{7.20}$$

where $\vec{r} = \vec{x}$, and D_r^2 represent the terms with $\partial/\partial r$. In the useful form one does *not* write out the terms in symbol $\vec{L}^2$, which is ideal in most simple form.

The usefulness happens because the eigenvectors of $\vec{L}^2$ are the spherical harmonics:

$$\vec{L}^2 Y_{\ell m} = \ell(\ell+1) Y_{\ell m};$$
$$L_z Y_{\ell m} = m Y_{\ell m}.$$

[12] The calculations using the collected formulas of many dozen authors have been done by John C Martens and the author.

Table 7.1. A collection of experimental uncertainties σ and data f_{expt} compared to our calculations which use two free parameters. The data for the $1S2S$ transition is the first item and has not been used in the fit.

σ_{expt}Hz	f_{expt} Hz	$f_{our\ calc}$ Hz
35	$2.46606141319 \times 10^{15}$	$2.46606141319 \times 10^{15}$
10074	$4.797\ 338 \times 10^{9}$	$4.79733066539 \times 10^{9}$
24014	6.490144×10^{9}	$6.49012898284 \times 10^{9}$
8477	$7.70649350012 \times 10^{14}$	$7.70649350016 \times 10^{14}$
8477	$7.7064950445 \times 10^{14}$	$7.70649504449 \times 10^{14}$
6396	$7.70649561584 \times 10^{14}$	$7.70649561578 \times 10^{14}$
9590	$7.99191710473 \times 10^{14}$	$7.99191710481 \times 10^{14}$
6953	$7.99191727404 \times 10^{14}$	$7.99191727409 \times 10^{14}$
12 860	$2.92274327868 \times 10^{15}$	$2.92274327867 \times 10^{15}$
20 568	4.197604×10^{9}	$4.19759919778 \times 10^{9}$
10 338	4.699099×10^{9}	$4.6991043085 \times 10^{9}$
14 926	4.664269×10^{9}	$4.66425337748 \times 10^{9}$
10 260	6.035373×10^{9}	$6.03538320383 \times 10^{9}$
11893	9.9112×10^{9}	$9.91119855042 \times 10^{9}$
8992	1.057845×10^{9}	$1.05784298986 \times 10^{9}$
20099	1.057862×10^{9}	$1.05784298986 \times 10^{9}$

The eigenvalues are real. Then automatically the $Y_{\ell m}$ are *orthogonal*, which makes them an ideal *basis* for representing functions of θ and ϕ. The orthogonality relations are

$$< Y_{\ell m} | Y_{\ell' m'} > = \int d\cos(\theta) d\phi \; Y^{*}_{\ell m}(\theta, \phi) Y_{\ell' m'}(\theta, \phi) = \delta_{\ell\ell'} \delta_{mm'},$$

where $-1 < \cos\theta < 1$ and $0 < \phi < 2\pi$ are the full ranges of the angles on the sphere. A general expansion is

$$f(\theta, \phi) = \sum_{\ell m} f_{\ell m} \, Y_{\ell m}(\theta, \phi); \qquad f_{\ell m} = < Y_{\ell m} | f > .$$

The $\vec{\nabla}^2$ operator appears in thousands of equations in mathematics, physics, engineering, chemistry, and so on. Every time spherical polar coordinates are used, one has the opportunity to replace $\vec{\nabla}^2$ by equation (7.20) in no steps. Next, one expands the function of interest in spherical harmonics, just like equation (7.17), with a possible addition of an index depending on m. Term by term where $\vec{\nabla}^2$ acts on the expansion one replaces $\vec{L}^2 \, Y_{\ell m} \to \ell(\ell + 1) Y_{\ell m}$. The angular derivatives disappear into the *eigenvalues*, which are just *numbers*, setting up subsequent steps in maximally easy form.

Because of these and many other nice properties, someone[13] has written that 'the spherical harmonics are better than hot home-made bread with butter on it'. The minimal information here is what needs to be mastered and used repeatedly. Unfortunately two distinct kinds of disinformation can possibly spoil the experience. First, if your source takes the not-very-useful form, invokes that thing called 'separation of variables', and launches into several pages about solving differential equations, it's almost a total waste of time, and also wrongheaded. The transformations start with equations you cannot solve, make arcane transformations, and end with equations named after dead mathematicians (associated Legendre) that you cannot solve. It is wrongheaded because its focus is on detail you don't generally want. It actually dates back a century ago to the time when basis functions and orthogonality were not understood, while memorizing differential equations was considered important, *and we're not doing that any more.*

Next, some sources might make an issue of the operator $\vec{L} = \vec{r} \times (-\imath \vec{\nabla}) = \vec{r} \times \vec{p}$ supposedly coming from the memorized substitution rules, and quantum physics. That was a common mistake of physicists with poor math training of (say) 75 years ago. There was a lot of guess work: 'we guess' $-\imath \vec{\nabla}$ stands for momentum, and 'we guess' $\vec{r} \times (-\imath \vec{\nabla})$ stands for 'angular momentum', and 'we guess' the $Y_{\ell m}$ will be involved in measuring angular momentum. The older presentations often mistook math facts to be physics discoveries. The mathematical consistency is automatic: the only physical information entering is the Schrödinger equation. While it was not recognized at first, in the 21st Century mature physicists know how Lie groups and generators work. (Someday you should learn about it.) There's no 'guessing' nor any reference to physical principles involved in the quantum theory of *angular momentum.*

Reference

[1] Schrödinger E 1926 *Ann. Phys.* **384** 361–77

[13] For once, Feynman did not say this.

Chapter 8

Observables

Figure 8.1. A rainbow trout photographed by Michael Stack of Fishtales Outfitting, mtfishtales.com.

The wave function represents our *description* of a physical system, in as much detail as our theory permits. As soon as a complete description of any complicated subject exists, there is usually a need to get rid of most of it, and work with simplifying summaries. *Observables* are numbers extracted from the wave function that can serve as simplifying summaries. This definition of 'observable' is absolutely neutral, and comes from examining a formula for a number, and interpreting it as a number.

doi:10.1088/978-1-6817-4226-7ch8

Yet 'observables' became a very contentious word through attempts to connect its definition with a very contentious 'doctrine of measurement'. We'll review the doctrine without forcing it upon you. Most of the turf-wars in early quantum mechanics were based on prescribing limiting and sometimes arbitrary restrictions on what would be 'allowed' as a good quantum measurement. It makes sense to review the general procedure for calculating observable numbers first, and after that consider additional restrictions for the cases when they may appear.

The question 'what is observable in quantum mechanics?' was raised well before quantum mechanics. The Bohr atom was built so that nothing about the atom itself was observable. As we mentioned in section 1.1.1, Born, Heisenberg, and Jordan (*BHJ*) at first promoted a philosophy of matrix mechanics based only on 'what was observable'. They did not define what that meant, and the operators used for calculations were *not* observable. When Schrödinger challenged everything with his wave function so easy to visualize, *BHJ* instantly dismissed it as 'unobservable'. In those days complex numbers were considered unobservable without a second thought. Then the whole history of the subject struggled with an apparent conflict between a perceived superstructure where calculations were made, and a different level where experimental predictions emerged. Whoever controlled the word 'observable' stood to command the subject, explaining the conflicts.

Physics evolved and moved on. We now understand we can, in principle, experimentally observe as much as desired that the wave function describes, and vice versa. While many things can be observed, there is a particularly appropriate way to integrate over the wave function with suitable weights, which makes a *number* that is very natural to describe experiments. It corresponds to the idea that one should make a correspondence between averaged properties of the wave function and things observed in the lab. These *collective* properties are both obvious and subtle at the same time. When you take for granted that the symbols correspond to the words, there's a second level of actually verifying that the interpretation is fair, faithful, and useful.

8.1 Collective position, velocity, and momentum: tropical storms

8.1.1 Collective velocity

Consider the weather map of figure 8.2. It has *too much* information. Weather scientists are then asked to give the 'the precise position' and 'the precise velocity' of hurricanes and tropical storms, as if they were point particles. There are no options but that a storm's position and velocity will be some kind of average summary, especially when the thing is 100 miles across.

Finding a good summary takes more than placing pins by eye on a satellite map. Whatever formula is chosen will have some arbitrary features, but they must be consistent. Suppose you want the collective velocity of a storm as a whole. Add up all the local velocity fluctuations over the storm's volume, which will cancel the whirling internal motions, and compute the total.

Figure 8.2. Exactly where is Hurricane Floyd, and where is he going? Quantities such as $<\vec{x}>$ and $<\vec{v}>$ provide convenient practical summaries of the behavior of quantum waves. NASA GOES satellite image.

Similarly, in quantum mechanics there exists a volume-integrated current denoted $<\vec{j}>$:

$$
\begin{aligned}
<\vec{j}> &= -\frac{\imath e_Q}{2\mu_Q}\int d^3x\ \psi^*\vec{\nabla}\psi - \psi\vec{\nabla}\psi^*, \\
&= -\frac{\imath e_Q}{\mu_Q}\int d^3x\ \psi^*\vec{\nabla}\psi.
\end{aligned}
$$

The second line comes from integrating by parts assuming $\psi \to 0$ at the boundaries of an arbitrarily large volume. Removing the charge parameter e_Q produces a candidate for a volume-integrated, collective wave velocity:

$$
<\vec{v}> = -\frac{\imath}{\mu_Q}\int d^3x\ \psi^*(\vec{x},t)\vec{\nabla}\psi(\vec{x},t). \tag{8.1}
$$

Often the time dependence of matter waves is artificially removed, and such formulas seem to have no sense of motion in them. That is why we made the movement explicit with symbol $\psi(\vec{x},t)$. The formula says that in some average sense the matter wave as a whole tends to move in an integrated way directed by $\vec{\nabla}\psi$, which may remind you of diffusion.

Equation (8.1) and almost all other uses of $\int d^3x = \int dxdydz$ instruct the research staff or a computer to add up a *number* over a volume, without reference to the much-despised operation of finding anti-derivatives in calculus courses. We'll repeat this early and often: the integrated quantities *exist* as symbols we can manipulate, while seldom needing a specific calculation. You should be just as neutral and unimpressed by it as in learning that the National Weather Service has people computing gradients of air pressure integrated over storm profiles. Also, keep in mind that totalized quantities are seldom specific enough to be very informative. If you know $<\vec{v}>$ you know three numbers, and cannot compute $\psi(\vec{x},t)$. Meanwhile if you know $\psi(\vec{x},t)$ it has the information of infinitely many numbers, and you can compute $<\vec{v}>$.

Given the wave function, many similar things can be calculated with a pointy bracket expression. The consistent definition is that <*something*> is a sandwich ψ^* *something* ψ, integrated over the whole volume:

$$< something > =< \psi|something|\psi > = \int d^3x\ \psi^*\ something(\vec{x})\ \psi. \tag{8.2}$$

The specific notation of $<\psi|\cdots|\psi>$ is used to distinguish one candidate ψ from another, when needed. Notice that $<\psi|\cdots|\psi>$ makes an 'operator sandwich' in which *operator* comes in the middle. It is necessary to choose an ordering for operators, which $<\psi|operator|\psi>$ resolves. One reason for the ordering is that $<\psi|operator|\psi>$ is real-valued whenever the operator is Hermitian. While that's not strictly necessary, nor needs to be a *postulate*, it is general enough to accomplish everything wanted.

It is not obvious, but $<-\imath\vec{\nabla}>$ is nearly unique to describe a collective velocity. There is no notion of $<d\vec{x}/dt>$, because $\vec{x}$ is not a function of time. Whatever you choose must be a true *vector* with the correct *transformation properties* of a velocity. For example $\int dx\psi(-\imath\vec{\nabla}>)\psi$ would not be real-valued. Taking the real part can be done, but that would not be *odd* under time-reversal. If you computed $<\vec{x}^2\vec{\nabla}>$ it would not be real, and would also depend on the coordinate origin for $\vec{x}$, while velocities do not. By mentioning time reversal and transformation properties, we're hinting there's a higher level of theoretical artistry and insight involved in relating operator sandwiches to physical quantities, which is true. Those who use *MSR* will not get everything wrong, but they will also not understand why.

The notation for pointy <> is borrowed from mathematics. For 100 years before quantum mechanics, mathematicians had noticed many isolated regularities of wave-like equations. The apparently stupid operation of integrating *function** × *something* × *function* consistently led to simple outcomes. Near 1900 the meaning of this was starting to clear up. By 1922 Hilbert and Courant published a book, *Methoden der mathematischen Physik* (Methods of Mathematical Physics) introducing *differential operators* to physicists who had no previous experience. Just as you'd expect, many physicists came to view $<\psi|\text{ANYTHING}|\psi>$ as something coming from the microscopic antics of inexplicable quantum *parcles*, rather than the direct fact it is just a *number*.

8.1.2 A concept error about momentum

We come to a concept error from early times. Consider equation (8.1). It is a *definition* that is not intended to be more than a crude summary. Now to make a mistake, we observe the dependence on $1/\mu_Q$ can be moved to the other side, giving

$$\mu_Q < \vec{v} > = -\imath \int d^3x \, \psi^*(\vec{x},t)\vec{\nabla}\psi(\vec{x},t). \tag{8.3}$$

In Newtonian physics, the formula for Newtonian momentum $\vec{p}_N = m_N\vec{v}_N$, where $v_N = d\vec{x}_N/dt$, and $\vec{x}_N(t)$ is the particle trajectory we said we'd never use. Yet at the dawn of quantum time a trajectory was desired, and the new meaning of quantum mass μ_Q was not recognized. That set up an interpretation error to claim (assert, discover, misunderstand) the right hand side of equation (8.3) must be the 'momentum of the quantum particle' we said we'd never use. The damaging part of the mistake asserts we would necessarily need to return to *parcles*, due to a strong unmentioned bias that 'if it has momentum, it's gotta be a particle'.

That is false since ocean waves and quantum waves and and tropical storms have momentum. The particle concept trying to pass the door does not belong, and makes no more sense than 'the hurricane particle' that destroyed a shopping mall[1] in Florida. *Also*, it makes no sense to invoke the Newtonian formula, which we're not using, to *define* a momentum for a wave, which is a completely different dynamical system. In any event, people were given false information that *momentum implies a particle*, and false information that particles define momentum, so that at the first mention of 'momentum' in quantum mechanics they were misled. It is no longer acceptable to make incoherent conceptual errors. Neither is it acceptable to make an unexplained statement that '$-\imath\vec{\nabla}$ IS THE MOMENTUM of a quantum *parcle* because it is a postulate'. That explains nothing, and we can do better than that.

8.1.3 Velocity second moment

The nth moment $\tilde{f}_n$ of a function $f(x)$ is a number from computing

$$\tilde{f}_n = \int dx \, x^n f(x).$$

The factor x^n weights large x regions more and more for large n. Suppose you are given f_n for $n = 1, \ldots, 30$. Then you know 30 numerical facts about the function, which is quite a bit of information. The *Mellin transform* of a function is just $\tilde{f}_n$ as a continuous function of n. Under very general conditions it is invertible: if you know all the moments of a function, you know the function. That is one example of how integral quantities like observables can probably encode all the information you might desire.

Suppose the first moment vanishes. Then the second moment $f_2 = \int dx \, x^2 f(x)$ is a measure of the squared width of the function. (Draw a picture of the integrand and assume $f(x)$ is well localized.) If $f(x)$ is normalized, a more general and standard

[1] Not to mention the effects of quantum hurricane particles on trailer parks.

definition of the function's width-squared is $\Delta x^2 = f_2 - f_1^2$. Subtracting f_1^2 makes Δx^2 not depend on the choice of coordinate origin, which is what we mean by the 'width relative to where the function is concentrated'.

The sandwich $<-\imath\vec{\nabla}>$ is actually a first moment of the spatial Fourier transform, namely the plane wave expansion of $\psi(\vec{x})$. This is seen from the expansion

$$\begin{aligned}\psi(\vec{x}) &= \int d^3k\ \tilde{\psi}(\vec{k})e^{\imath\vec{k}\cdot\vec{x}};\\ -\imath\vec{\nabla}\psi(\vec{x}) &= \int d^3k\ \vec{k}\tilde{\psi}(\vec{k})e^{\imath\vec{k}\cdot\vec{x}}.\end{aligned} \tag{8.4}$$

This uses the fact that each plane wave $e^{\imath\vec{k}\cdot\vec{x}}$ is an eigenvector of $-\imath\vec{\nabla}$ with eigenvalue $\vec{k}$. A common notation writes

$$\vec{K} \equiv -\imath\vec{\nabla}; \qquad \vec{K}|\vec{k}> = \vec{k}|\vec{k}>.$$

Using equation (8.4) in the sandwich and reversing the order of integration gives

$$\begin{aligned}<\vec{K}> &= \int d^3k \int d^3x\ \underline{\psi^*(\vec{x})e^{\imath\vec{k}\cdot\vec{x}}} \times \vec{k}\tilde{\psi}(\vec{k});\\ <\vec{K}> &= \int d^3k\ \underline{\tilde{\psi}^*(\vec{k})}\vec{k}\tilde{\psi}(\vec{k}),\end{aligned} \tag{8.5}$$

$$\text{where} \int d^3x\ \psi^*(\vec{x})e^{\imath\vec{k}\cdot\vec{x}} = \tilde{\psi}^*(\vec{x}).$$

Some terms were underlined to make inspection easier. These expressions ignore (make invisible) the $\sqrt{2\pi}$ 'normalization clutter' of conventional Fourier transforms[2]. Our objective is equation (8.5), which shows that $<\vec{K}>$ produces a vector of first moments of $\tilde{\psi}^*(\vec{k})\tilde{\psi}(\vec{k})$. If a different operator 'A' were in the sandwich, expanding in the basis of its eigenstates would make a similar relation for the moment of the operator's eigenvalues 'a'. This is formalized in section 9.1.3.

We are ready for the second moment $<\vec{K}^2>$. In no steps compute

$$<\vec{K}^2> = \int d^3k\ \tilde{\psi}^*(\vec{k})\vec{k}^2\tilde{\psi}(\vec{k}).$$

This suggests a candidate for the second moment of velocity $<\vec{v}^2>$. By agreement we define the *velocity operator* to be $\vec{K}/\mu_Q$, and then $<\vec{v}^2> = <\vec{K}^2>/\mu_Q^2$. Notice that $<\vec{v}^2> \neq <\vec{v}>^2$, so this is new information. When acting on $\psi(\vec{x})$ the squared-velocity operator may not be transparent: we agree to use the plane wave eigenstates where $\vec{k}^2$ is simple for the definition.

Let's emphasize we are exploring *agreements* to assign words to calculations. This is much different (and much more honest) than a tradition pretending some mysterious conspiracy by words would predict the physics. As it stands, $<\vec{v}^2>$ is 'one more numerical fact' about the wave function. We actually know more:

[2] Replace $dx_i \to dx_i/\sqrt{2\pi}$ and $dk_i \to dk_i/\sqrt{2\pi}$ component to make normalization factors explicit.

$\Delta K^2 = <\vec{K}^2> - <\vec{K}>^2$ measures the squared-width of the wave number components of the wave function.

Suppose $<\vec{v}> = 0$, so the wave as a whole is vibrating, but nominally (by naming it) the quantum system is 'at rest'. From the Schrödinger equation

$$\begin{aligned} \frac{1}{2\mu_Q}\vec{K}^2\psi &= \imath\partial\psi/\partial t - U(\vec{x})\psi, \\ \frac{1}{2}\mu_Q < \vec{v}^2 > &+ < U(\vec{x}) > = < \Omega >, \end{aligned} \tag{8.6}$$

where Ω and $\imath\partial\psi/\partial t$ are equal acting on solutions to the Schrödinger equation.

We mentioned it was often thought that $U(\vec{x})$ was *always* the Newtonian potential energy function. To do some guessing, irresponsibly replace $\frac{1}{2}\mu_Q < \vec{v}^2 > \to \frac{1}{2}\mu_Q < \vec{v}>^2$, and one term looks like Newtonian kinetic energy. The other term is some average (that depends on ψ) of the potential energy. When Ehrenfest was finding his way with the symbols, the relation seemed 'almost' the sum of kinetic and potential energies of the Newtonian system he had in mind, suggesting Ω would be the *energy operator*, which had been claimed. From those analogies $<\vec{K}^2>/2\mu_Q = \mu_Q < \vec{v}^2 > /2$ is generally called the 'average kinetic energy' of a quantum system. That is a misnomer, since $\psi(\vec{x})\vec{\nabla}^2\psi$ actually measures potential energy of bending the wave, but it is seldom noticed.

As it stands, $<\Omega>$ *does not involve* $\mu_Q < \vec{v}>^2/2$, so the Newtonian matchup does not quite work out. Yet despite this apparent flaw, we can show that $<\Omega>$ is a *constant* in time, just as needed for a *conserved energy*. The proof is quite easy, using the Schrödinger equation:

$$\begin{aligned} \frac{\partial}{\partial t} < \Omega > &= < \frac{\partial\psi}{\partial t}|\Omega|\psi > + <\psi|\Omega|\frac{\partial|\psi>}{\partial t}>, \\ &= <\psi|(\imath\Omega)\Omega|\psi> + <\psi|\Omega(-\imath\Omega)|\psi> = <\Omega^2 - \Omega^2> = 0. \end{aligned} \tag{8.7}$$

Like any conservation law, this is a useful fact about the wave. But what does it mean?

The old way of guessing relations using the *MSR* was certainly confusing, and succeeded or failed about 50% of the time. That's why we recommend more modern methods. There's no reason for guessing when Hamiltonian physics is well-defined on its own. An unambiguous calculation (section 8.2.1) shows that $<\Omega>$ is *exactly* the energy of the system. This does not come from *Newtonian* physics or analogies, but from the straightforward facts of continuum mechanics.

The existence of a strictly conserved energy contradicts fanciful interpretations of wave packets having 'no definite energy' because of the 'uncertainty principle'. It is true that a wave packet with no single frequency has no single frequency. The guess $E = h\nu$ was *never* a correct definition of energy, which is *always* a quantity describing an entire system. There are actually no cases where the uncertainty *relation* predicts information about a wave function. If you have any wave function whatsoever, it already contains more information than the little scrap of fact from the uncertainty relation. If you know this, you will escape many fallacious references to the uncertainty relation.

8.1.4 Collective position

The formula for $<\vec{v}>$ can be related to a formula for the *average position* of the hurricane, storm, or quantum wave. It is not trivial to define a competent measure of average position. The typical average position of 'stuff' is a weighted average of the stuff multiplied by its position. The electron wave stuff is represented by $\psi(\vec{x},t)$, but what kind of average shall we choose? Here are three candidates[3] for an average position:

$$<\vec{x}>_a \stackrel{?}{=} \int d^3x\ \vec{x}\psi(\vec{x},t);$$
$$<\vec{x}>_b \stackrel{?}{=} \int d^3x\ \vec{x}|\psi(\vec{x},t)|;$$
$$<\vec{x}>_c \stackrel{?}{=} \int d^3x\ \psi^*(\vec{x},t)\ \vec{x}\ \psi(\vec{x},t).$$

The first line is complex, disturbing some people, so avoid it. The second line uses $|\psi(\vec{x},t)| = \sqrt{|\psi^*(\vec{x},t)\psi(\vec{x},t)|}$, which signals an error of math inexperience: you cannot compute anything after a square-root-abs operation[4]. For one thing, $|\psi(\vec{x},t)|$ will have cusps at zeroes of ψ, making a mess. The last line may not be a work of art, but it's better than the first two candidates. *The whole of the discussion has its arbitrary features*: the habit (of the old school) that by making crude summaries, we are discussing something deep about existence *just stinks.*

In favor of the third definition, Ehrenfest was able to show from the Schrödinger equation that

$$\frac{d}{dt} <\vec{x}>_c = <\vec{v}>.$$

To see it is not trivial, put $<\vec{v}>$ on the left side and pose the relation as a question:

$$\begin{aligned} <\vec{v}> &= -\frac{\imath}{\mu_Q} \int d^3x\ \psi^*(\vec{x},t)\vec{\nabla}\psi(\vec{x},t), \\ &\stackrel{?}{=} \int d^3x\ \frac{\partial\psi^*}{\partial t}(\vec{x},t)\ \vec{x}\ \psi(\vec{x},t) + \psi^*(\vec{x},t)\ \vec{x}\frac{\partial\psi(\vec{x},t)}{\partial t}. \end{aligned} \tag{8.8}$$

The equation is not a monster, just look at both sides. The time dependence is in $\psi(\vec{x},t)$, the moving wave. It ends up as time dependence $d<\vec{x}>/dt$, an averaged substitute for a trajectory. The only way the relation can be true is if $\imath\partial\psi/\partial t$ is related to derivatives such as $\vec{\nabla}^2\psi$ by a wave equation of motion. Use the free-space Schrödinger wave equation:

[3] We introduce the 'is it equal' sign $\stackrel{?}{=}$ used on physics blackboards everywhere, perhaps appearing here for the first time in printed form.

[4] Math has easy and hard directions, like wood. It's a pleasure to discover the easy directions, and avoid going across the grain that will punish you.

$$\iota \frac{\partial \psi}{\partial t} = -\frac{1}{2\mu_Q} \vec{\nabla}^2 \psi. \tag{8.9}$$

Replacing $\partial \psi / \partial t$ gives

$$-\frac{\iota}{\mu_Q} \int d^3x\, \psi^* \vec{\nabla} \psi \stackrel{?}{=} -\frac{\iota}{2\mu_Q} \int d^3x\, \vec{\nabla}^2 \psi^*\, \vec{x}\, \psi - \psi^*\, \vec{x}\, \vec{\nabla}^2 \psi. \tag{8.10}$$

Notice that the ordering of operations must be respected. *Operator ordering* is not new with quantum mechanics, but many calculations need a new level of attention to it. The first term on the right hand side is related to the second by *FIP*:

$$\int d^3x\, (\vec{\nabla} A) B = -\int d^3x\, A \vec{\nabla} B \leftarrow \text{FIP};$$

$$\int d^3x \left(\vec{\nabla}^2 A\right) B = -\int d^3x\, (\vec{\nabla} A) \cdot \vec{\nabla} B = \int d^3x\, A \vec{\nabla}^2 B \leftarrow \text{FIP}$$

FIP uses $\int d^3x \vec{\nabla}(integrand) \to 0$ when the integrand vanishes at the boundaries, which is always assumed by using boundaries so far from the system that *FIP* applies[5]. Using *FIP* with equation (8.10) gives

$$-\frac{1}{\mu_Q} \int d^3x\, \psi^* \vec{\nabla} \psi \stackrel{?}{=} -\frac{\iota}{2\mu_Q} \int d^3x\, \psi^* \left(\vec{\nabla}^2 \vec{x} - \vec{x}\, \vec{\nabla}^2\right) \psi.$$

For the x component calculus gives

$$\psi^* \left(\frac{\partial^2}{\partial x^2} x - x \frac{\partial^2}{\partial x^2} \right) \psi = 2\psi^* \frac{\partial \psi}{\partial x}. \tag{8.11}$$

That gives the x component of the left side of the equation above. The same follows for the other components, creating Ehrenfest theorem-1.

FIP is often learned first in quantum mechanics. It is formalized in terms of a Hermitian operator. By *FIP* $<\vec{\nabla}>^* = -<\vec{\nabla}>$, because ψ and ψ^* trade places. Introduce a factor of ι and then

$$<\vec{K}>^* = <-\iota \vec{\nabla}>^* = <-\iota \vec{\nabla}> = <\vec{K}>,$$

or more simply, $<\vec{K}>$ is automatically real. As we said often, any operator H such that $<H>$ is real for any ψ is called Hermitian. To repeat,

$$H \text{ is Hermitian if and only if } <\psi|H|\psi> = \text{real for all } \psi.$$

If you resist knowing this, don't. Consider the convenience of working with real-valued summaries, so that you don't need to think about complex numbers.

A person knowing *FIP* might work backwards and guess the Schrödinger equation from equation (8.8). In fact the Ehrenfest theorem-1 works *for any real-valued interaction*

[5] *FIP* is not a cheat. It replaces an unnecessary 'postulate' about the Universe with the option to recompute any cases where *FIP* might not apply.

function $U(\vec{x})$. In equation (8.10) replace $-\vec{\nabla}^2\psi/2\mu_Q \rightarrow -\vec{\nabla}^2\psi/2\mu_Q + U(\vec{x})\psi$, and nothing changes. The proof of the continuity equation is similar, and also needs $U(\vec{x})$ to be real.

8.1.5 The expected classical limit that failed

The objective of the 'Ehrenfest relations' was to obtain Newtonian physics as an output, where the quantum theory would predict the classical trajectory $\vec{x}_N(t) \overset{?}{\rightarrow} \vec{x}_Q = <\vec{x}>(t)$, and then recover Newtonian dynamics. We have seen part one, which shows two definitions of $<\vec{x}>$ and $<\vec{v}>$ are compatible. Making definitions compatible is competent mathematics, not a fact of nature. To get his objective to come out, Ehrenfest would need to predict the Newtonian formula for 'force' from the acceleration $m_N d < \vec{v}>/dt$. This did not work out, as can be seen with no work. The calculation of $d < \vec{v}>/dt$ will depend on the wave function $\psi(\vec{x},t)$ defining $<\vec{v}>$:

$$\frac{\partial}{\partial t} < \vec{v}> = -\frac{\imath}{\mu_Q}\frac{\partial}{\partial t} \int d^3x\, \psi^*(\vec{x},t)\vec{\nabla}\psi(\vec{x},t).$$

The right hand side depends on ψ, no matter what substitutions are made. Yet the Newtonian force is a given and fixed formula that does not depend on $\psi(\vec{x},t)$. The two cannot be equal.

Amazingly, one textbook after another ignored this. The equations were manipulated to give $<\partial U/\partial x>$ on the right hand side. With a statement that the interaction function was 'the classical potential energy'—not from logic or experiment, but by *postulate*—the sandwich $<-\partial U/\partial x>$ was called the Newtonian force. While the formula is not the Newtonian force, scholars still wrote that the 'classical limit' had been recovered. For example, Griffiths [1] writes that '(the equations) are instances of Ehrenfests's theorem that expectation values obey classical laws'. Those were the expected words, but the expectation value the passage cites is *not* the classical law[6].

The mistake was caused by the presentation flow chart invented by Bohr. It wrongly inserted particles and particle probabilities as defining elements in all of the equations and before defining anything else. That made ignoring the flaws of Ehrenfest mandatory! The uncertainty relation was claimed early and out of place, making everything downstream seem to be inexact. The symbols $<\vec{x}>$ and $<\vec{v}>$ were presented as statistical averages (expectation values), which all agree can *sometimes* be motivated. By fouling up the logical order the Ehrenfest relations were made to seem 'good enough,' especially when the order of averaging was too murky for a student to challenge it.

Yet there is a conceptual error in treating exact relations as probability relations. Statistical averages have fluctuations, while exact relations do not. As we've just seen, the symbols $<\vec{x}>$ and $<\vec{v}>$ and $<\partial U/\partial x>$ are *exact* quantities. The

[6] Not all books are wrong. The book by *Commins* makes a series expansion to identify *special circumstances* where the expected Newtonian answer would apply. We don't know what happens otherwise! The textbook by *Ballentine* has a mature discussion of serious concerns.

contradictions of a wrong interpretation cannot be ignored. If a person said: 'Wait! The Newtonian force does not ever depend on the wave function', his teachers would be hard-pressed to respond[7]. Prematurely introducing quantum probability both botched many definitions, and also put everything derived downstream from it on intellectually unstable, and sometimes *incompetent* grounds. It's the road we're not following, and for good reasons!

8.2 The general definition of observables

Our investigation of $<\vec{x}>$, $<-\imath\vec{\nabla}>$, etc, used sandwiches which are 'quadratic', namely bilinear in $\psi^{*}...\psi$. For the general definition:

$$\begin{aligned} &\text{A quantum observable} < H > \text{of the wave function } \psi \\ &\text{is a number obtained by} < H > = \frac{<\psi|H|\psi>}{<\psi \mid \psi>}, \\ &= <\psi|H|\psi> \text{ assuming } <\psi|\psi> = 1. \end{aligned} \tag{8.12}$$

It has usually been assumed that H is Hermitian so the observable is real. If H is not Hermitian, one gets a complex number, equivalent to two real numbers. With combinations of non-Hermitian sandwiches real numbers can be made to come out: it is a meaningless 'postulate' to *require* Hermitian operators. To avoid fighting with those who have not thought this through, we'll use Hermitian operators.

The *non-observability* of $<\psi|\psi>$ is something new. It does not agree with one's notion of a classical continuum wave. It can be explained: the wave function is an eigenvector of an operator, and eigenvectors never have a definite, predetermined normalization. Explaining more here would be a distraction: there is more information in section 10.3.3. One will usually see $<\psi|\psi> \to 1$ imposed as a 'normalization postulate'. The definition dividing out $<\psi|\psi>$ avoids an unnecessary postulate and shows that $<\psi|\psi> \to 1$ is a *normalization convention* that is convenient.

Any multiplicative or linear differential operator can be put into a sandwich, to learn some information. For example $<x^3>$ provides a number coming from the wave function not available from $<x^2>$ and $<x>$, so it is new information. $<x^4>$ provides another number. Students will ask: 'What is the correct quantity to use?' Nobody in authority can tell you. The actual situation in physics is that observations need to be matched to experiments, and this is never trivial. Any numbers providing information will be welcome: the idea they could be universally specified in advance is too far-fetched to consider. *However*, you will find a stream of literature suggesting that $\vec{x}$ and $-\imath\vec{\nabla}$ observables more or less complete 'all possibilities', inasmuch as it's what a *parcle*-person would think about. We turn to a few interesting topics.

[7] The typical response would be: 'Ehrenfest becomes exact as $\hbar \to 0$'. That's also false, since $\hbar$ has not played any role, but the bias to believe in the magic of $\hbar$ was very strong.

8.2.1 Collective wave momentum

WE REGRET THE FACT quantum mechanics met a few unsavory efforts to systematically divert the origin of math relations into physical postulates about the Universe. At first it seems harmless: who's to decide what postulate is more fundamental than any other? Yet the harm done is considerable. It can be entirely avoided by careful ordering of presentation.

For example, we presented that equation (8.3) is only tenuously connected to 'momentum'. By honest steps we showed the expression is a computable 'sandwich' of $\psi^* \ldots \psi$ with $-\imath\vec{\nabla}$ in between, and not more was claimed. In the most dogmatic approach a 'physical postulate' was made that $-\imath\vec{\nabla}$ 'IS! IS!! IS!!! the momentum of the quantum particle'. Depending on your source, it was often posed to *deliberately* be without context, origin, or any possibility of understanding it. It was also *very importantly* claimed to be independent, inexplicable, fundamental: hence, impossible to challenge. Instead of talking about a physically moving wave, the symbol $-\imath\vec{\nabla}$ was set up as THE OPERATOR FOR MOMENTUM, WHERE IN QUANTUM MECHANICS, MOMENTUM IS AN OPERATOR.

One does not expect physics to be a struggle over controlling word usage, but that's what happened early on. First, $-\imath\vec{\nabla}$ IS NOT Newtonian momentum, which has a different use and definition. Next, if $-\imath\vec{\nabla}$ is used *to calculate* some kind of momentum, we don't need the overbearing IS! IS!! IS!!! existence nonsense. Right?

Now as we mentioned, waves of all kinds, including the quantum kind, do carry momentum. The Schrödinger equation predicts everything there is to know about the dynamics. The facts and definitions of momentum come from dynamics, and are found inside the rules, not outside. A competent physicist should be able to derive the formula for wave momentum from the wave equation.

The most simple derivation goes as follows. In general form the Schrödinger equation is written

$$\imath\frac{\partial\psi}{\partial t} = \Omega\psi,$$

where often $\Omega \to -\vec{\nabla}^2/2\mu_Q + U(\vec{x})$. The corresponding action is

$$S = \int dt\mathcal{L}(\psi, \psi^*) = \int dt d^3x \; \imath\psi^*\frac{\partial\psi}{\partial t} - \psi^*\Omega\psi. \tag{8.13}$$

The Schrödinger equation is found from the extrema $\delta S = 0$. The second term is $<\Omega>$. The Lagrangian above has the general form $L = \sum_i p_i\dot{q}_i - H(q_i, p_i)$, showing that $<\Omega>$ IS the Hamiltonian functional.

The momentum of any generalized coordinate q is $\partial\mathcal{L}/\partial\dot{q}$, where $\mathcal{L}$ is the Lagrangian, and $\dot{q}$ is the time derivative. Consider a wave of any shape $\psi_{\text{any}}(\vec{x})$ whose time dependence comes from a simple translation by coordinate $\vec{q}(t)$. Thus

$$\psi_{\text{any}}(\vec{x},t) = \psi_{\text{any}}(\vec{x}-\vec{q}(t)).\leftarrow\text{translation}.$$

The vector $\vec{q}(t)$ represents the collective coordinate for translations, which make only three variables among the formally infinite wave coordinates. Under the translation shown, the collective position changes

$$
\begin{aligned}
<\vec{x}> \to <\vec{x}>(t) &= \int d^3x\, \psi^*_{\rm any}(\vec{x} - \vec{q}(t))\vec{x}\psi_{\rm any}(\vec{x} - \vec{q}(t)), \\
&= \int d^3x'\, \psi^*_{\rm any}(\vec{x}')(\vec{x}+\vec{q}(t))\psi_{\rm any}(\vec{x}'), \\
&= <\vec{x}>+<1> \vec{q}(t) = <\vec{x}>+\vec{q}(t).
\end{aligned}
$$

The last line divides by the normalization $<1> = \int d^3x\psi^*\psi$, or uses a normalized wave function. The result transforms like the position on a map, just as we expect (and insist) for the words 'collective position'.

The calculation of the collective momentum conjugate to coordinate $\vec{q}$ is

$$
\begin{aligned}
\vec{p}_q &= \frac{\partial \mathcal{L}}{\partial \dot{\vec{q}}} = -\imath \frac{\partial}{\partial \dot{\vec{q}}} \int d^3x\, \psi^*_{\rm any}(\vec{x}-\vec{q}(t))\frac{d}{dt}\psi_{\rm any}(\vec{x}-\vec{q}(t)) \\
\text{where } &\frac{d}{dt}\psi_{\rm any}(\vec{x}-\vec{q}(t)) = -\dot{\vec{q}}\cdot\vec{\nabla}\psi_{\rm any}; \\
\text{then } \vec{p}_q &= -\imath\frac{\partial}{\partial\dot{\vec{q}}}\dot{\vec{q}}\cdot\int d^3x\, \psi^*_{\rm any}\vec{\nabla}\psi_{\rm any}, \\
\text{and then } \vec{p}_q &= -\imath\int d^3x\, \psi^*_{\rm any}\vec{\nabla}\psi_{\rm any}.
\end{aligned}
\tag{8.14}
$$

More generally, *Noether's theorem*, published in 1918, establishes the *local* character of conservation laws in continuum mechanics. Symmetry under spatial and time translations leads to conservation of the *energy momentum tensor*, for example, which Noether's theorem predicts. The basic relation between conservation and symmetry is much older in Lagrangian and Hamiltonian physics. Noether's approach shows how to systematically formulate those particular generalized momenta conjugates to particular coordinate variations. Our application using a collective coordinate builds a bridge between beginner's (few coordinates) and more advanced (infinitely many coordinate) Lagrangian physics.

A Lagrangian formulation is necessary because all general concepts of *momentum* and *energy* are Lagrangian–Hamiltonian concepts, which cannot be formulated correctly otherwise, as we said earlier. It is a bit unfortunate that this particular Lagrangian material comes from a higher level than our discussion. Sorry, energy and momentum come from Lagrangians, and can't be dumbed-down. If you don't know about Lagrangians, you must learn about them!

8.2.2 Constants of the motion

In section 3.2.2 we mentioned a tendency for Hamiltonian physics to be misunderstood. *Not understanding* the subject meets the agenda of the *N*-people and the *C*-people, where *N* and *C* stand for Newton and Copenhagen. It's strange but true that the *N*-people tend to dislike Hamiltonians, because the method makes classical physics too easy. The beautiful *relation between symmetry and conservation laws* in Hamiltonian physics takes no steps. The Hamiltonian equation $dp_i/dt = -\partial H/\partial q_i$ says that if $H(q_i, p_i)$ does not depend on any generalized coordinate q_*, the corresponding p_* is constant in time. No steps!

Since quantum wave dynamics is of the classical Hamiltonian type (while involving gazillions of times more variables) the relation between symmetry and conservation holds true. It happens to look different when expressed with operator sandwich observables. Compute the time dependence of an observable using the Schrödinger equation:

$$\begin{aligned}\frac{d}{dt} < A> &= \frac{\partial}{\partial t} < A > =<\frac{\partial \psi}{\partial t}|A|\psi > +<\psi|A|\frac{\partial|\psi}{\partial t} >, \\ &= < \psi|(\imath\Omega)A|\psi > +<\psi|A(-\imath\Omega)|\psi > =\imath < \Omega A - A\Omega >.\end{aligned} \tag{8.15}$$

The algebra assumes the operator is not explicitly time dependent; otherwise add $\partial A/\partial t$. On the right hand side we see the *commutator* $[\Omega, A] = \Omega A - A\Omega$. With that notation we have

$$-\imath\frac{d}{dt} < A > =<[\Omega, A] >. \tag{8.16}$$

This repeats the calculation done for equation (8.7). In no steps, $<A>$ is constant in time if $[\Omega, A] = 0$. But what does it mean?

It is the obvious and beautiful relation of conservation and symmetry, to those skilled in the art of reading it. Commutators are discovered and developed in the theory of Lie groups. The meaning of a commutator $[A, B]$ is the rate of change of B under the transformation generated by A. The concept of 'a transformation generated by A' comes from Lie groups. Let $\alpha \to 0$ be an infinitesimal real parameter. Consider the transformation operator $\mathcal{U}(\alpha) = 1 + \imath\alpha A$. One says that '$A$ generates the transformation $\mathcal{U}(\alpha)$'. Acting on a state gives

$$|\psi > \to|\psi >_\alpha = \mathcal{U}(\alpha)|\psi > =(1 + \imath\alpha A)|\psi >.$$

If $|\psi>$ does not change under the transformation, there is a *symmetry* under the operator $\mathcal{U}(\alpha)$. Also, $A|\psi > =0$ must be true. Then $<[\Omega, A] > = 0$, and $<A>$ is a constant of the motion.

This algebra so far is a little incomplete. The word *symmetry* refers to a feature of a physical *system*, and only sometimes a feature of a state. The transformation $\mathcal{U}(\alpha)$ can be considered a change of basis of the entire system. Under the transformation the Hamiltonian matrix will change by $\Omega \to \Omega_\alpha = \mathcal{U}(\alpha)\Omega\mathcal{U}^\dagger(\alpha)$. Either you know this from linear algebra, or you must learn it. By algebra

$$\begin{aligned}\mathcal{U}(\alpha)\Omega\mathcal{U}^\dagger(\alpha) &= (1 + \imath\alpha A)\Omega(1 - \imath\alpha A), \\ &= \Omega + \imath\alpha[A, \Omega] + O(\alpha^2),\end{aligned} \tag{8.17}$$

and assuming $A = A^\dagger$ (as always) for observables. This equation shows that the change of Ω_α per parameter α is the *commutator* with A, which generates the transformation. When $[A, \Omega] = 0$ then to first order[8] in α the transformation leaves

[8] The requirements of a *group* turn out to be so demanding that many relations shown to first order in a small parameter such as α will be true to all orders. $[\Omega, A] = 0$ is sufficient to show Ω is unchanged under a certain 'finite transformation' generated by A with an arbitrarily large parameter α.

Ω unchanged, and that is a *symmetry*. If you are following this skill of the art, then in no steps equation (8.16) says that $<A>$ is constant when Ω has a symmetry generated by A, expressing the relation between symmetry and conservation.

By the way, the Hamiltonian Ω is *the generator of time evolution*, because $\Omega \rightarrow \imath\partial/\partial t$ every time it acts. The *time translation operator* $\mathcal{U}(t) = 1 - \imath t\Omega$ creates a change of basis moving an infinitesimal duration t into the future.

Example. Free space has translational symmetry. That means Ω is not an explicit function of $\vec{x}$, and $[-\imath\vec{\nabla},\Omega] = [\vec{K},\Omega] = 0$. In no steps $<\vec{K}>$ is constant. Also, $\vec{K}$ and Ω share eigenstates, namely $e^{\imath\vec{k}\cdot\vec{x}}$, which helps solve them.

Discussion. A sort of foggy reverence for quantum mechanics, and corresponding disdain for classical Hamiltonians, has tended to present these results a little differently. In the early days of quantum theory those who knew all about Lie groups had mathematical super-powers, and seemed to have been ungenerous in sharing their powers. There was then (and now) a definite tendency to present all of the above as a miraculous outcome of quantum physics. That is not logical: everything is coming from the Schrödinger equation, which is a generic wave equation, with all the rest being mere *notation*. Putting factors of Planck's constant everywhere gave people false information that commutators were *zero* when $\hbar \rightarrow 0$. That's absurd, given that Lie groups and commutators were known to come from mathematics, and to represent the underlying geometry of transformations, more than 30 years before the physicists found commutators in quantum notation. In effect, somebody early on lied to other people about commutators. Next, a series of guesswork analogies associated with the *MSR* (section 5.3.2) actively asserted that the operator methods were quasi-Hamiltonian *by analogy* and by guesswork, either through ignorance or by concealing the construction of the actual and literal Hamiltonian of the theory (section 8.2.1). Relations like equation (8.16) were presented as marvelous accidents of guesswork, not the direct outcome of *notation*, which is the real explanation.

Our telegraphic review of Lie groups cannot begin to do justice to the subject. There's obviously much to be developed. Whether or not you completely absorb what we've described, it is *very important* you have the information that those methods exist, and *a physicist must eventually learn all of them*. Those in the 21st Century who know enough about Lie groups continue to have super-powers over those who don't!

8.2.3 Observability of the wave function

Suppose $\psi(x) \rightarrow \psi_x$ is pre-evaluated in $N \gg 1$ slots of x. If $N = 1000$ is enough for you, choose it. The are N^2 possible combinations $\psi_x\psi^*_{x'}$, including the 'diagonals' $x = x'$ and 'off diagonals' $x \neq x'$. The sandwich of any operator $H^{(a)}$ cannot be more complicated than evaluating an $N \times N$ matrix sum:

$$<\psi|H^{(a)}|\psi> = \sum_{x}^{N} \sum_{x'}^{N} \psi_x^* H_{xx'}^{(a)} \psi_{x'}.$$

Consider the slot $x = x' = 1$. Let $H_{11}^{(a)} = 1$, and all other $H_{xx'}^{a)} = 0$. Then $<H^{(a)}>$ reports the $\psi_1\psi_1^*$ value. Let $H_{12}^{(b)} = 1$, and all other values be zero. Then $<H^{(b)}>$ reports the $\psi_1\psi_2^*$ value[9]. Continue by choosing operators $H^{(a)}$, $H^{(b)}$... $H^{(N^2)}$ all terms in ψ can be evaluated. So the wave function is observable.

There's one loophole. Multiplying $\psi \to \lambda e^{i\theta}\psi$ cancels out every time: one cannot detect it with an operator sandwich. But we have already agreed the overall normalization is not observable. The observable part of the wave function predicts what is observed and vice versa.

This is the gist of Fano's proof [2] using *density matrices* that with enough measurements-as-sandwich observables the wave function is observable. It surprises and shocks many who assumed (and were told) the wave function could not be observed. There's a couple of reasons not to be overly impressed. *What are all these $H^{(a)}$, $H^{(b)}$...$H^{(N^2)}$ things physically?* Math will not tell you. They stand as a mathematical concept correction to the mathematical concept error that the wave function *could not possibly be* observable. *How will one choose $H^{(a)}$, $H^{(b)}$...$H^{(N^2)}$ in an experiment?* Math will not tell you. An impression from the old stuff that operators must be 'position' $\vec{x}$ or 'momentum' $-i\vec{\nabla}$ is confronted: the habitual narrowness of discussing nothing more than Newtonian variables in Copenhagen bites the dust. *How could you make infinitely many measurements?* We're sure we cannot. But do you need infinite information? Why? Moreover, many useful quantum systems are described by rather few variables, like (say) three numbers. Can you measure three numbers? The upshot is that one catastrophic early conceptual error—'complex numbers are imaginary and cannot physically exist'—did a great deal of early damage. Once corrected, the ordinary supposition that *you can measure just as much as you have the time, patience, and facilities to measure* is most of the message.

One more thing: since determining the wave function well might take a hundred or a hundred thousand measurements (billions at a particle accelerator), it would be illogical to assign your final wave function to every single individual system. You are not finding out features of individual systems, but finding out features of collections, often called 'ensembles' of individual systems. *The wave function is a descriptive tool of ensembles.* By awfully clever math, to emulate the ensemble of little waves, you emulate one element of a little wave, and read out the predictions using the rules of quantum mechanics.

Einstein is said to have opposed quantum mechanics because he opposed a probability interpretation. That is false. Einstein opposed irrational dogmatism. In 1935 Einstein, Podolksy, and Rosen wrote the single most important paper in

[9] The perceptive reader will notice we're not insisting H is Hermitian. As we noted, it is redundant: it assumed, the math is the same, because $\psi_x\psi_{x'}^*$ is automatically Hermitian, which decreases the number of independent measurements by about half.

quantum mechanics, after Schrödinger's 1926 paper, according to many. We recommend you not to believe internet physa-bloggers who repeat that 'God does not play dice' yet never read Einstein's papers and letters. In 1949 he wrote[10]:

> 'The attempt to conceive the quantum-theoretical description as the complete description of the individual systems leads to unnatural theoretical interpretations, which become immediately unnecessary if one accepts the interpretation that the description refers to ensembles of systems and not to individual systems.'

These are not the words of a person stubbornly opposing probability. The word 'ensemble' accepts a statistical interpretation. Einstein's remarks are totally correct. There is universal evidence that quantum mechanics describes ensembles of systems. Like the photon, Einstein's beliefs were misreported and misrepresented. There is a mystery here, to discover how and why Einstein's character was assassinated in his own lifetime. Who benefited?

8.2.4 The uncertainty relation: never say 'principle!'

Did you know that the uncertainty *relation* is not a *principle*. Principles have priority, predictive power, and cannot be derived from math. Relations are just math facts, and are always derived. It's nice that *Wikipedia* recognizes it:

> 'Historically, the uncertainty principle has been confused [4, 5] with a somewhat similar effect in physics, called the observer effect, which notes that measurements of certain systems cannot be made without affecting the systems, that is, without changing something in a system. Heisenberg offered such an observer effect at the quantum level (see below) as a physical 'explanation' of quantum uncertainty [6]. It has since become clear, however, that the uncertainty principle is inherent in the properties of all wave-like systems [7], and that it arises in quantum mechanics simply due to the matter wave nature of all quantum objects.'

For once *Wikipedia* has it right, except the word 'principle' should be corrected to 'relation'. Any wave of any sort obeys the relation *automatically* and without referring to any authority of a 'principle'.

In 1927 Heisenberg noticed an interesting fact about wave blobs, which was not obvious in over-simple cosine waves. He was experimenting with Schrödinger's theory, which he previously called 'crap', precisely because it was easy to understand. Heisenberg noticed the math fact we'll soon describe, and sent his draft paper

[10] The quote is found in [3]. The source cited is [4].

(or a letter, accounts differ) to Bohr. At that point Bohr convinced[11] Heisenberg that the uncertainty principle was a manifestation of the deeper concept of complementarity. Heisenberg duly appended a note to this effect to his paper on the uncertainty principle, before its publication, stating:

> 'Bohr has brought to my attention [that] the uncertainty in our observation does not arise exclusively from the occurrence of discontinuities, but is tied directly to the demand that we ascribe *equal validity* to the quite different experiments which show up in the [particulate] theory on one hand, and in the wave theory on the other hand.'

The note has a bizarre reference to 'the occurrence of discontinuities', and was intended to give those concepts (reiterating the *OQT*) a validity that did not exist. No calculations of the actual quantum theory supported 'discontinuities', so they appeared in a note added. Also note our italics, 'equal validity'. Instead of getting rid of the *parcle* that was not there, a *parcle* reappeared in a note added, the best way to get equal validity! That was Bohr's idea of complementarity (which few remember now) that was so big back then. The 'occurrence of discontinuities' also came from Bohr's belief system. Like the particle, it was nowhere found in the math relation we're talking about.

You will understand the meaning, the inputs and the outputs when you do the calculation. Today anyone with a computer can check it. Figure 8.3, top left panel, shows a series of values for a bell-shaped function $f(x) \rightarrow f_x$ where the index $0 \leq x \leq 40$ in steps of one unit. The actual wave is continuous, while it is represented as a list of values, with one value per slot labeled x. (This is not only how the math works—whether or not the label x is continuous—but also how we bypass integral expressions with their potential math-intimidation factor.) You have now seen the input: a function. We are not doing physics at this point, and there are no physical assumptions.

The Fourier series corresponding to the function is a weighted sum of waves going like $\tilde{f}_k \cos(kx)$ and $\tilde{f}_k \sin(kx)$ that adds up to give the original function[12]. The series coefficients $\tilde{f}_k$ are not 'outputs', because they are simply a representation of the input. (Recall the trick of no information, equation (5.4).) The notation is that the amplitude (weight) for each $\cos(kx)$ is the value of $\tilde{f}_k$. The bottom left panel of Figure 8.3 shows the $\tilde{f}_k$ corresponding to $f(x)$, case (a). The signs of $\tilde{f}_k$ alternate in this example, so the summed-up cosine waves will correctly cancel each other well to make the bell-shaped function $f(x)$, and that's quite a stunt. The relationship of the original formula and coefficients is an exact formula, executed by the computer in milliseconds.

[11] This passage comes from [5].

[12] Fourier series use *real* parts for even functions, cosines, and *imaginary* parts for odd functions, sines. It's another example of 'two parts' of complex numbers, with nothing 'imaginary' about it.

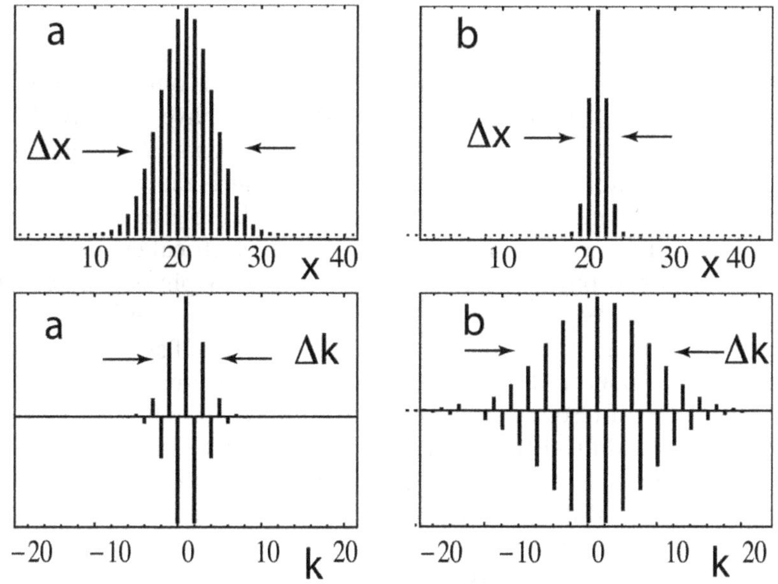

Figure 8.3. The uncertainty relation is a math fact. The left panels (a) show a typical function $f(x)$ sampled at points which makes a list f_x (top), and its Fourier expansion coefficients $\tilde{f}_k$ (bottom). Large width Δx makes a small range Δk. The right panels (b) show the same for a more localized function. Small width Δx makes a large range Δk. There is a math fact that $\Delta x \Delta k \geq 1/2$.

Coming to the point, let Δx and Δk be reasonable ranges of important x and k components. Notice that (case (a), left) large width Δx makes a small range Δk. Compare the right side panels, case (b). Here a more localized function with smaller Δx makes a larger range Δk. Since Δx and Δx are in an inverse relation, a math fact exists:

$$\Delta x \Delta k \geq \frac{1}{2}.$$

The 'proof' of such a fact consists of giving natural and precise definitions of the quantities Δx and Δk. In one dimension the standard choice is

$$\Delta x^2 \equiv <x^2> - <x>^2 = <(x - <x>)^2>,$$

with a similar expression for Δk. It is more important to understand that Δx and Δk are simple order-of-magnitude estimators than belabor the derivation.

This concludes the content of the uncertainty *relation* (*ur*). When $\Delta k \to 0$ then $\Delta x \to \infty$, which expresses the fact that a cosine wave with a perfect repeating wavelength must be infinitely long, else contradict itself. Conversely, *representing* a somewhat localized wave-blob by a weighted sum of cosines needs quite a spread of wavelengths to cancel everywhere except on the blob. Then $\Delta x \to 0$ and $\Delta k \to \infty$. Now if you are suspecting that the *ur* is not that interesting after all, you actually do have the right picture. The complications mathematicians and physicists face when expressing their own arbitrary functions as sums of other arbitrary functions cannot be called a crisis in anyone's world-view[13].

Did the analysis produce new information? No. The *ur* is only an inequality, used for rough estimates. Better estimates exist. In fact, if you have any wave function in front of you, the full Fourier expansion already exists, representing all the information not available from the inequality. So the *ur* is the kind of mathematics that *decreases* the total information going into it, in order to give a simple and useful fact about waves. That is how the relation is used in electrical engineering, plasma physics, oceanic wave physics, etc, as well as quantum physics. Since time and frequency f are related by a Fourier transform, there is also a *ur* written $\Delta f \Delta t \geq 1/2$, as anticipated in section 2.2.3. (Sorry for two different uses of symbol f.) If you want an instrument to respond to time fluctuations as short as a nanosecond (10^{-9} s), its *bandwidth* or spread of resolved frequencies must exceed about $10^9 \mathrm{s}^{-1}$, or more. The *ur* is more about the difficultly of building sharp shapes or rapid fluctuations with smooth waves than it is about localization. A wave with a kink may be quite delocalized, yet will typically obey the *ur* inequality by having a much wider range of Δk than one might suspect.

Unfortunately, some older works present the *ur* as 'physics' associated with the eigenvalue postulate. According to the postulate, every measurement of position x must be infinitely sharp: no errors! It is uncompromising and literal, $\Delta x = 0$, which needs $\Delta k \rightarrow \infty$, which needs $< \Omega > \rightarrow \infty$, namely, infinite energy. The premises are simply unphysical. Watch out also for mixups based on putting probability claims too early. Quite a few educated people cannot answer the question: 'Is $\Delta x \Delta k \geq 1/2$ true always, or only true in an average sense?'. Some will hedge that 'it's always true, *on average*', which equivocates to hide the fact statistics might be lucky. To avoid the confusion, keep in mind that exact math facts cannot fail. From its derivation the 'uncertainty' relation $\Delta x \Delta k \geq 1/2$ is an exact math fact that is true by the definition of the symbols. It is a *certainty* relation.

8.3 Logjam restrictions on observables

There is a different use of the word 'observable', which should have been called 'restricted observable', and which by omitting the qualification caused great controversy and confusion. It comes from the 'eigenstate postulate'.

The issue comes from *some very special experiments* which rather clearly filter out an eigenvector of a physically relevant operator. For example, pulling on electrons emerging from a hot filament with an electric field, and letting them escape into free space, rather easily produces a beam well enough described by $e^{i\vec{k}\cdot\vec{x}}$. If the beam is messy and has many components of $\vec{k}$, the crafty experimentalist will tinker with it, because a pure and simple beam is such a nice tool. Let $|\vec{k}>$ stand for an arbitrarily narrow wave packet, and $\vec{K}$ stand for $-\imath\vec{\nabla}$ operating on it. Since $-\imath\vec{\nabla}e^{\imath\vec{k}\cdot\vec{x}} = \vec{k}e^{\imath\vec{k}\cdot\vec{x}}$ is a true eigenvalue equation, it can be written

[13] The key step was choosing the word 'uncertainty'. If it had been called the 'blob-relation' it would not be famous.

$$\vec{K}|\vec{k}> = \vec{k}|\vec{k}>.$$

The observable using $\vec{K}$ is

$$<\vec{K}> = \frac{\vec{k}<\vec{k}|\vec{k}>}{<\vec{k}|\vec{k}>} = \vec{k}.$$

Notice this is not profound. To repeat:

IF YOUR STATE HAPPENS TO BE AN EIGENSTATE OF AN OPERATOR,
THE OBSERVABLE IS THE CORRESPONDING EIGENVALUE.

Now comes the converse. The ‘eigenstate postulate’ states that *every quantum measurement yields the eigenvalue of a Hermitian operator and the corresponding eigenstate*. The claim is *devoid of information* but also *circularly true when it happens to be true.* The postulate is not actually a part of quantum mechanics, *except among those who insist it be a part*, and it is not an experimentally supported postulate. There are many ways to evade it: even when an experimental apparatus has been tuned up to filter out eigenstates, you can add up the observations in ways not equalling eigenvalues. For example, accepting the premise that $\vec{k}$ acts like Newtonian momentum, the momentum of 10^{20} electrons in a beam very poorly made with mixed up wave packets can be delivered to a block of wood, which recoils, and no eigenstates are ‘measured’. For another example, the eigenstate postulate (*ep*) says that when you measure an electron spin, you will find one of two eigenstates called ‘up’ or ‘down’. (And confusing ‘what’s claimed you will measure’ with conditions that are inherently up or down is very common, as encouraged by the *OQT*.) Yet in a medical MRI device the continuous range of electron polarizations of the electron spins is the *signal* which is actually *observed.* The breakdown of the *ep* only happens when real-world experiments are studied. There is one place it has never failed: *the eigenstate postulate is universally true in schoolbook thought experiments that assume it holds.*

Someone might be offended when we call the *ep* meaningless[14], but we don’t know why. The perception there is content came from a mindset focused on making postulates *before* there was a Schrödinger equation. Before there was a quantum theory a handful of classic experiments more or less spontaneously exhibited outcomes that seem to need the *ep*. The 1923 Stern–Gerlach experiment is always cited. It is an outcome of the wave function property called *entanglement* (section 10.3). As a consequence it explained without the *ep after* you have a wave function. (And whatever state $|\tilde{\psi}>$ comes out, there exist a Hermitian operator $|\tilde{\psi}><\tilde{\psi}|$ that makes the *ep* circularly true, which is rarely noticed.)

To some extent the *ep* attempts to cover the gap of the *BHJ* program that had operators and no wave function. Since there was no state, there was no notion of *<anything>*, and no way of calculating what the (‘observable-based’) program would observe. Making ‘observables’ identical to eigenvalues closes the gap with an operator-based definition. It is also as close as one can possibly come to perpetuating

[14] Preposterous is also good: it means ‘contrary to reason or common sense; utterly absurd or ridiculous’. It’s not an insult, but the right word for those wanting quantum mechanics not to be understandable.

the *OQT* with its 'intrinsic quantization'. For example, we have seen that $<\Omega>$ is a number equal to the energy of the system. The energy is *not* generally quantized, whether or not you have a bound state. An authority appears, and dictates that 'you will measure an eigenvalue', and then (by a juggle of words) the 'energy is quantized when you measure it'. Those words—it is not quantized in a superposition, but it is quantized when you measure it—are an example of the confusion inherent in the *OQT* and its supporters. It backtracks to the Bohr model, and on the same authority basis of *arbitrarily postulating what was wanted to come out.* Directing attention to eigenvalues comes as close as possible to intrinsic quantization while avoiding the wave function, the initial conditions, and even the time evolution that are the actual accomplishments of quantum mechanics.

The *ep* comes with a 'collapse postulate' many have noticed is inexplicable, and fishy. The statement is that 'the measurement of the eigenvalue causes the wave function to collapse to the eigenstate', where (usually) it is claimed to persist until the next measurement. One should notice that if your apparatus does select an eigenstate, the observation of an eigenvalue and an eigenstate are circular, nothing is collapsing, and it does not need a principle. More generally the eigenstate and collapse (also called 'projection') postulated omit the crucial interaction with a measuring instrument that would be involved in detecting the wave function. It was omitted early because the theorists did not know what was involved, and omitted again by the authority figures who enjoyed making postulates. The crux of the issue is whether the wave function describes an *ensemble* of experiments (as a statistical proxy) or whether it defines a separate 'reality' for individual systems. In the first case there's no crisis when your *description* collapses or your wave function evolves: your information is a different thing from the 'Universe'. The second case, however, has generally been asserted, along with an overblown role for the wave function, which has caused a problem for reality when reality instantly collapses. We don't really understand the psychological appeal this has for some people.

The *persistence* by postulate of an eigenstate under time evolution must also be noticed. If it is true, the operator observed must commute with the Hamiltonian (section 8.2.2). But then steady wave functions, namely frequency eigenstates, are *automatically* eigenstates of the commuting operator. Why does one need a postulate to predict what's already known from math?

The other reason for the *ep* is to set up highly idealized sequences of measurements that dramatically contradict the assumptions of pre-quantum physics. The drama is enhanced by the theorem of linear algebra that two operators A, B automatically have joint (common) eigenstates if they commute, $[A, B] = 0$. They cannot have common eigenstates if $[A, B] \neq 0$. In that case, when a 'proper' measurement yields an eigenstate of A, it cannot have been a 'proper' measurement of B. Since this is an obvious deductive statement from math, there's no new physical information in it. However, an early contingent wanted to think that $-\imath\vec{\nabla}$ IS! momentum, and $\vec{x}$ is! position, so then $[\vec{x}, -\imath\vec{\nabla}] \neq 0$ made 'simultaneously measuring the position and momentum of the quantum particle' impossible. (And we agree!) The other purpose of the *ep* is to make measuring the wave function impossible. Suppose the wave function has N complex degrees of freedom, or 'dimension N'. It will be described by $2(N-1)$ real numbers, since one overall complex normalization is not used. There are exactly N Hermitian operators that commute, including 1. (They can all be diagonal, filling N diagonal slots.) There are not as many *commuting* operators as variables in the wave function, shutting the door on observing it *under the added restrictions of the eigenstate postulate.*

The *fact* that commuting operators have joint eigenstates is not a postulate, but a very useful *fact*, often confused with the *ep*. For example, any interaction function with rotational symmetry commutes with the angular momentum operators $\vec{L}^2$ and $\vec{L}_z$, (section 7.2.2) which commute with each other. Since the eigenfunctions of $\vec{L}^2$ and $\vec{L}_z$ have been known for 150 years, the joint eigenstate property instantly reduces spherically symmetric problems to much easier problems (section 7.2.2). This can be misunderstood by blindly starting with the *ep* pretending to be a universal fact, mistakenly believing the system must exist solely in those commuting eigenstates, and misinterpreting the formulas... so don't do that!

Summary of issues with the eigenvalue postulate Let's summarize:

- An observable in an operator sandwich is a number like $<A>$.
- If the state is an eigenstate, $|\psi> = |a_n>$ with $A|a_n> = a_n|a_n>$, then $<A> = a_n$. The normalization $<\psi|\psi>$ cancels out, or choose $<\psi|\psi> = 1$ by convention.
- With enough observables one can formally reconstruct the wave function. That, by the way, may be quite ambitious.
- While not necessary in quantum mechanics, a contingent maintains every observable IS an eigenvalue. It is a Law of Nature of Schoolbooks few experimental physicists ever take seriously. This different use of 'observable' cannot be defended, and so is made a 'postulate' of the Copenhagen presentation.
- The eigenstate postulate restricts 'proper' measurements to operators which share eigenstates, and then must commute. This is a circularly true fact of so-called 'proper' measurements.
- The *restrictive* measurements insisting on eigenstates are not enough to determine $|\psi>$. And so what?
- The *ep* causes a common confusion that what 'exists' might be those *restrictive* measurements and eigenvalues. It often succeeds in replanting the mistakes of the *OQT*: And that is really bad for people!

It's not commonly known that Einstein thought the attempt to restrict nature by authoritarian prescriptions was ridiculous. When someone suggested one of his thought experiments might violate the restriction of the eigenvalue postulate, he replied[15]: 'I couldn't care less' ('Ist mir wurst').

[15] Letter of Einstein to Schrödinger, 19 June 1935.

8.3.1 FIAQ

Is there something wrong with the uncertainty relation or principle here? Heisenberg's relation involves $\hbar$. The math relation has no $\hbar$. Which one is the true relation? They are the same relation. We asked you to keep track of assumptions. The wave-derived relation has no role and no room for Planck's constant. To put it in, Heisenberg used *algebra* and the *OQT* that a *parcle's* Newtonian momentum $\Delta p_N = \hbar \Delta k$. The algebra was

$$\Delta x \Delta k \geq \frac{1}{2},$$

$$\Delta x(\hbar \Delta k) \geq \frac{\hbar}{2}.$$

In almost all of math and physics, an equation multiplied on both sides by the same constant is the same equation. The uncertainty *principle* is the first time in history that a math identity became a physical principle by multiplication on both sides by a constant that cancels out.

It is an interesting experiment to ask people who know a little quantum physics about this. Some of them will say you cannot divide out $\hbar$. ('That will break a math rule.') Some will say that when you do such a division, you change the meaning of the equation. But if that's the case, will the meaning change if we multiply all terms by 17.3? The people victimized by the *OQT* were so bamboozled by $\hbar$ they said it deserves a special dispensation to sit in equations and provide extra meaning, which would cancel out for any constant *except* $\hbar$.

Some may remain 'uncertain'. Bohr and Heisenberg jointly worked up a 'photon microscope' presentation somewhat above the level of high school physics. It purports to describe an interaction: there is none. The microscope argument introduces a notion of *momentum*, just where the *parcle* enters. The wave relation $\Delta x \Delta k \geq \frac{1}{2}$ is not obtained from *momentum*, so some believe an extra meaning might seem to come from $\hbar$ and the classical limit $\hbar \to 0$. Here's the secret: waves carry *momentum*. It's unhelpful, and in the post-quantum presentation, incompetent or deceptive to introduce momentum with a fictitious *parcle*. Since momentum was not needed for the math inequality, it would be incompetent or deceptive to introduce it at all. We have already developed wave momentum from the Schrödinger equation and its Lagrangian (section 8.2.1). It stands on its own.

References

[1] Griffiths D 2005 *Introduction to Quantum Mechanics* 2nd edn (Upper Saddle River, NJ: Prentice Hall) p 18

[2] Fano U 1957 *Rev. Mod. Phys* **29** 74

[3] Ballentine L 1998 *Quantum Mechanics: A Modern Development* (Singapore: World Scientific)

[4] Einstein A 1959 *Albert Einstein: Philosopher–Scientist* ed P A Schlipp (New York: Harper & Row)

[5] Mehta L and Rechenberg H 2000 *The Historical Development of Quantum Theory* vol VI (New York: Springer)

John P Ralston

Chapter 9

More ways to describe waves

Figure 9.1. Amazing waves in the sky of Lääne-Viru County, Estonia, 2014, by Kairo Kiitsak.

9.1 More than one description

We've developed material with some depth. Let's step back to simpler things, and clear up a few concepts that often get muddled.

9.1.1 Two ways to add waves

First, without any physics, or anyone's permission, you can verify that adding two functions makes a new function. If someone wants to add cosine shapes of various wavelengths, you cannot stop them. If someone (Mr Fourier) claims that almost any shape can be reproduced from a sufficiently complicated sum of sine and cosine shapes, it's a math fact, which is self-defining. (It will work out, for exactly where it works out.) There is no statement about nature in any of that.

For example, suppose you are observing a wave in a plastic bucket of water. In your mind you consider the peak of the wave to be the sum of two smaller waves. You consider each smaller peak to be the sum of smaller ones, and the smaller peaks to be made of smaller ones, all the way down to tiny ripples. Once you've entertained

doi:10.1088/978-1-6817-4226-7ch9

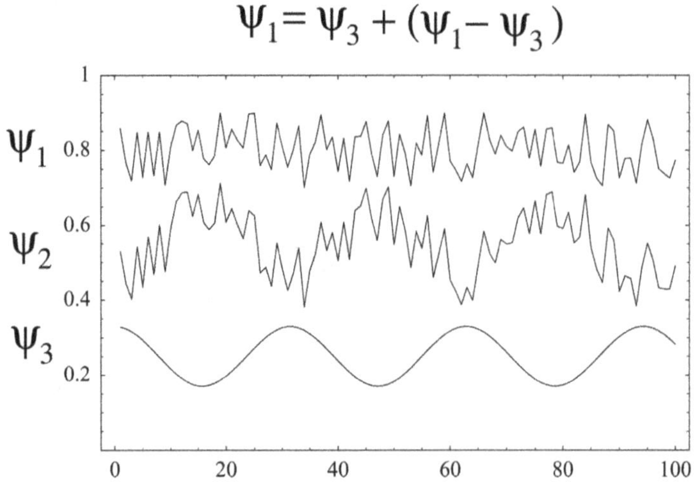

Figure 9.2. By an amazing equation, *any* wave (top) *is absolutely equal* to any other wave (bottom) plus the difference (middle). It's a hoax to say that this comes from a quantum principle, and it's a swindle to say that the waves are 'in every state at once'.

yourself, you tell people that your bucket is in many states at the same time. It is both one big wave, it is also two smaller waves, and it is 10 000 ripples of myriad shapes, all at once, because that's how you want to think about it. By an amazing math fact (figure 9.2), *absolutely any* wave *is absolutely equal* to any other wave… plus the difference added as a third wave. (And this is trivial, right?)

Next and independent of this, *a certain class* of wave equations, which are called 'linear', have the feature that you can make a genuinely new solution by adding any weighted sum of old solutions. That is remarkable. It makes solving equations so easy it is like cheating. But let's not cheat, and call the single class where a property applied a 'principle'. If an equation is non-linear, you cannot add solutions. Yet writing a solution $A = B + (A - B)$ can always be done. No one can stop you from expanding any solution in an orthonormal basis. So there is actually two kinds of superposition, consistently confused by those who use the term *'principle of superposition'* cited early, in section 3.1.1. We recommend burning the words 'principle of superposition' in a bundle of western sage from a new age bookshop, and never using it again.

9.1.2 Superposition is not a quantum effect

When quantum mechanics is developed, one learns that waves of different shapes can be viewed as vectors on a high-dimensional space. This is quite a thrill, which seems wonderfully abstract. In the early confusion about quantum mechanics, most physicists, and many mathematicians, did not understand how that kind of math worked. Out of that era came stories about an electron being in all states at once, which is a misinterpretation of the mathematics. From that era came wrong statements that 'superposition is a uniquely quantum mechanical effect'.

Figure 9.3 shows how any vector can be considered to be the sum of two other vectors. The figure actually shows two different ways, among infinite ways, to make a sum. This fact is a very confusing thing for people learning vectors of Newtonian force.

Whether the vectors represent forces or wave functions, the idea of a vector sum is the same. The possibility of representing a Newtonian force vector an arbitrary number of ways was never considered a big philosophical issue. But if you wish to consider an electron to be in 'all states at once', you might by symmetry consider the Newtonian analog with just the same basis. Make and label as many 'force triangles' as you desire with the letters of your choice. Considering all these alternatives, you might consider saying the classical Newtonian forces are in every direction, and no particular directions, all once. But you will not say it, because you will see it is both a concept error and an abuse of language, just as for the quantum case.

The first mistake in saying that 'superposition is a uniquely quantum thing' is forgetting to count the number of dynamical variables. The quantum *parcle* concept thinks about three variables, and cannot understand all the wave coordinates, which must be used. Next, *almost all* representations of physical quantities use some superposition of coordinate labels. The labels of the wave function are actually on the same footing as Newtonian phase space coordinates, with a significant difference. The Newtonian coordinates are not pinned down very well by the theory: they are pinned down by the user. NASA uses a very precisely defined celestial coordinate system for the navigation of spacecraft. That coordinate system is ultimately defined by a number of distant stars. In quantum mechanics the wave function is expressed in *implicitly defined* coordinates, which are not pinned down so well. You can take all the eigenfunctions of the hydrogen atom, and make an arbitrary unitary transformation, and still describe a hydrogen atom. The standard Schrödinger equation has been posed in a coordinate convention that makes it easy to visualize, but *nature is not concerned with our coordinate conventions.* Moreover, a physical hydrogen atom will be interacting with the Universe in ways we do not control, and *that is where* the natural expressions of its coordinates will appear, much like the stars fixing a convention for the astronomers.

9.1.3 The eigenstate expansion of observabies

Consider an operator A, with eigenvalues a_n and eigenstates $|a_n>$:

$$A|a_n> = a_n|a_n> \tag{9.1}$$

Suppose $|\psi>$ is not a eigenstate of A, and you measure $<\psi|A|\psi>$, which will not equal an eigenvalue of A. To avoid the work of computing integrals, one expands $|\psi>$ in the basis of eigenvectors of A:

$$|\psi> = \sum_n \psi_n|a_n>,$$

$$\text{where}\quad \psi_n = <a_n|\psi>; \qquad A|a_n> = a_n|a_n>$$

$$A|\psi> = \sum_n \psi_n a_n|a_n>.$$

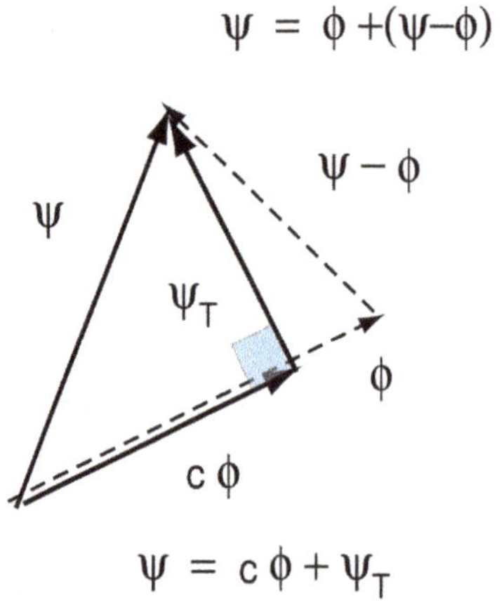

Figure 9.3. By an amazing equation, *any* wave (top) *is absolutely equal* to any other wave (bottom) plus the difference (middle). The basic equation is $A = B + (A - B)$. It's a hoax to say that this comes from a quantum principle, and it's a swindle to say that the waves are 'in every state at once.'

Remembering that $<\psi|$ stands for ψ^*, its expansion is

$$<\psi| = \sum_m \psi_m^* < a_m|.$$

We used a new index m in the sum: it is very common not to remember this, so watch what comes next. Combining the formulas gives

$$< \psi|A|\psi > = \sum_m \sum_n \psi_m^* < a_m|a_n > a_n\psi_n = \sum_n a_n\psi_n^*\psi_n = \sum_n a_n| < a_n|\psi > |^2. \quad (9.2)$$

The ugly double sum collapses because orthogonality gives zero for almost all terms except the matching cases $m = n$. Once again the normalization $< a_m|a_n > = \delta_{mn}$ greatly simplifies things. If an index like n is used in two different sums, it causes a 'global variable' inconsistency that will confuse your calculations, which is very common. The last term in the equation above is redundant, added to make the notation self-explanatory.

Inspecting equation (9.2), the trick of using basis $|a_n >$ reduced the operator A to its eigenvalues, and nothing could be more simple. Pay attention to this! It's very ingenious. Next, the expression is real if a_n are real, which proves the theorem defining Hermitian operators. If you do not know or prove all theorems of linear algebra before coming to quantum mechanics, basic calculations will discover all theorems.

Finally the expression provides a clue for those cases where one would believe the eigenstate postulate. It claimed every proper measurement gave an eigenvalue. Equation (9.2) is a sum of eigenvalues weighted by $\psi_n^*\psi_n$, which is consistent with $\psi_n^*\psi_n$ being the *probability to find each eigenvalue*. Consistency is one thing: does it really happen?

We can easily imagine some real-world situations. A number of experiments are done, each one happens to give an eigenvalue, and the probability of each case is counted up by the frequency of occurrence. This natural experimental process happens *sometimes.* It is different from knowing $|\psi>$ from enough measurements, and computing $<\psi|A|\psi>$ directly, perhaps never getting any lucky eigenstates. Then *in the event* the eigenstate postulate holds, there seems to be little room to interpret $|\psi_n|^2 = | <a_n|\psi> |^2$ as anything but the probability to find state $|a_n>$, given $|\psi>$. The range $0 \leq | <a_n|\psi> |^2 \leq 1$ is certainly consistent.

Most variants of the eigenvalue postulate include another postulate that eigenvalues must occur with probability $| <a_n|\psi> |^2$. Our presentation exposes a weakness: it's not the only possibility. In fact one should not define 'probability' by the frequency of events. Probability is a deep and subtle subject. It is a sort of artificial intelligence of managing and manipulating information, whether or not it is uncertain information. The most powerful and interesting developments of *Bayesian* probability explicitly avoid associating probability with event frequencies.

It is best to limit interpretations to what stands on its own. As constructed, one can say that $| < a_n|\psi> |^2$ is so far *consistent with a type of probability* to find state $|a_n>$, given $|\psi>$. Since probability represents a mathematical structure, more exploration will be needed to determine exactly what kind of probability this might be.

9.1.4 Heisenberg picture

We briefly mentioned earlier the *time evolution operator* $\mathcal{U}_t$, which is a beautiful and powerful concept such that $\psi(t) = \mathcal{U}_t\psi(0)$. We claim that from the Schrödinger equation $\mathcal{U}_t \sim 1 - \imath t\Omega$ for time parameter $t \to 0$. (Check this is the same as $\imath\, \psi/\partial t = \Omega\psi$.) For those skilled in the art of Lie groups (which we do not expect) an operator using the same symbol but for arbitrarily large t will *solve* the time evolution by a coordinate transformation $\psi(t) = \mathcal{U}_t\psi(0)$. The full time evolution operator is an awesome thing, which can only be constructed for a few solvable models, but whose *existence* unites many advanced topics.

With that existence in mind, we 'compute' (actually represent) the time dependence of observables by

$$< A > (t) = < \psi(t)|A|\psi(t) > = < \psi(0)|\mathcal{U}_t^\dagger A\mathcal{U}_t|\psi(0) > .$$

This allows one to *define* a time-dependent operator

$$A(t) \equiv \mathcal{U}_t^\dagger A\mathcal{U}_t. \tag{9.3}$$

That makes $< A > (t) = < \psi(0)|A(t)|\psi(0) >$. With the time dependence in the operator, the expression uses $\psi(0)$, which is the initial condition frozen at the initial time $t = 0$. That defines the 'Heisenberg picture': use time dependent operators, constant $\psi(0)$, and calculate all the same numbers. For example, the operator $\vec{K} = -\imath\vec{\nabla}$ can be mapped in a time dependent operator $\vec{K}(t)$, which created notation reminiscent of Newtonian physics. Yet far from providing any feature of a particle trajectory, the time dependent operator is an object of 'infinity-squared' difficulty. The dimension of any operator is the dimension of the wave function, squared.

Since the time dependence of $A(t)$ is known, a differential equation exists to predict it. It is called the Heisenberg equations of motion, although Heisenberg's original (1925) equation was not exactly the same, and the derivation we are following came from Schrödinger. Those equations are generally *non-linear*, whenever the Hamiltonian operator is not a simple quadratic function of other operators, which is usual. Even worse than usual, the fact of being non-linear means there is no hope for analytic solutions. To this day there exist essentially no independent solutions of the Heisenberg equations, except solutions to solved Schrödinger equation problems that are mapped into time dependent operators.

If one asks why any of this is relevant, it can be explained. First, the solved case of a quadratic Hamiltonian ('harmonic oscillator') is extremely useful because the operators have an infinity-squared amount of information. The mathematical power is used to set up practical approximation methods. Next, the early competition between Born, Heisenberg, and Jordan versus Schrödinger kept the operator-based approach on center stage for priority, while for a while appearing to be a distinctly different approach to quantum mechanics. However, one notices that if the time dependent operators had any dynamical existence as physical entities, the equation of motion would not be so trivial as producing equation (9.3). In general non-linear operator equations would be *much more complicated* and come in tremendous variety. It is not noticed because Schrödinger coyly wrote that the two approaches of matrix mechanics and wave mechanics had been shown to be 'equivalent'. It was an overstatement that minimized embarrassment to *BHJ*. Those words 'equivalent' have been repeated everywhere without recognizing the *BHJ* program lacked the essential concept of a quantum mechanical *state* with its initial conditions, which Schrödinger provided.

Even before Schrödinger had published, Cornelius Lanczos in 1925 recognized that the abstract, difficult, non-linear operator equations of *BHJ* matrix mechanics were reproduced by an infinitely more simple *linear integral* equation. To a skilled mathematician like Lanczos, that was a signal that the non-linear operator equations represented essentially an error in formalism that made an easy thing infinitely complicated. This accurate criticism is almost never seen, while it is useful information to balance a somewhat mystical reverence sometimes given to the approach. Yet Lanczos was ignored: he was not in mainstream physics, he was in 'no-stream' physics. Lanczos had missed the element of *communicating* with physicists. Legend has it that he visited Pauli, the most corrosively negative (while brilliant) physicist of the time. Pauli is said to have dismissed Lanczos as worthless, ignorant, hopeless. Yet Lanczos was not a fool: he published his observations before Schrödinger. However Lanczos missed the even easier *linear differential* form of Schrödinger's theory, and he did not solve any useful problems. If Lanczos had persisted and reproduced the hydrogen atom spectrum, then *Lanczos* would have been the hero of quantum mechanics. The integral form of the equation was rediscovered, and exists in a highly useful form as the *Lippmann–Schwinger equation*. Unlike the Heisenberg picture, nobody considers it an independent approach to quantum mechanics, but instead an excellent mathematical method beautifully suited to certain applications.

9.1.5 Comparing waves

There are essentially two ways to compare the shape of waves, or functions in general. One 'boneheaded fussy' way considers any function to be 'completely different' from another unless both agree everywhere. One function-as-a-list $f_x = (f_1, f_2, \ldots, f_n)$ is different from another if any element of the list is different. This is the bookkeeping method of factory part numbers and librarians. Decimal numbers are lists with a convention sorted by powers of 10. It is a much bigger mistake to confuse 31 190 with 81 193 compared to 31 198, but if a list is different, it is anyway and forever *mutually exclusive* in the boneheaded fussy method.

Yet wave-shape-lists that increase, decrease, or oscillate can be equally important. They have no natural order. It is also not reasonable to consider a shape differing by a small bump or wiggle to be completely different from another one. A very efficient and elegant comparison of shapes comes from the inner product, also called the *overlap*. We have been using it to some extent: the inner product is so important that repetition and redundancy is generally needed. We will deliberately let this section be self-contained, and hope you recognize things seen before.

In visual terms, two reasonably localized functions $f_a(x)$ and $g(x)$ are graphed simultaneously. The product $f_a(x)g(x)$ is large where both $f(x)$ and $g(x)$ are large. If the functions are well separated then one is zero while the other is large, and vice versa. Please look at figure 9.4. The *overlap* denoted $<f|g>$ of two real functions is computed from the area of the product $f(x)g(x)$ everywhere the functions exist:

$$
\begin{aligned}
<f|g> &= \int_{-\infty}^{\infty} dx\, f(x)g(x) \quad \leftarrow \text{overlap of real } f,\, g \\
&\sim \left(f_1 g_1 + f_2 g_2 + \cdots f_n g_n\right) \Delta x.
\end{aligned}
\tag{9.4}
$$

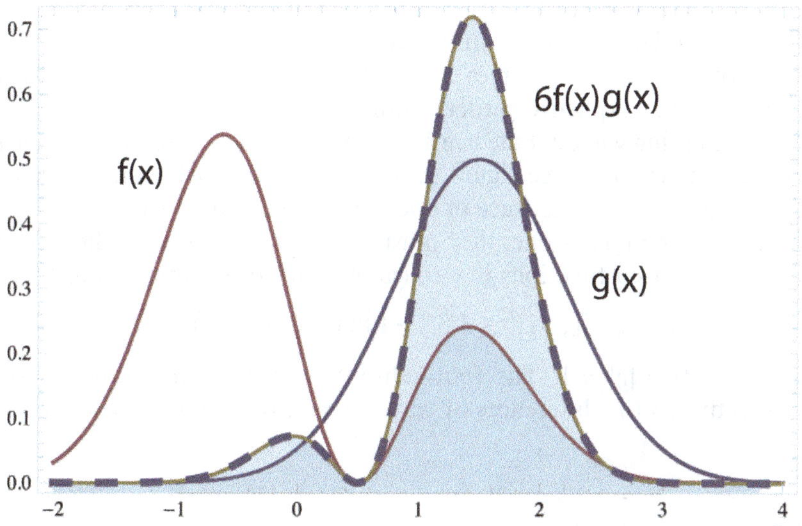

Figure 9.4. To compute the *overlap* of two functions $f(x)$, $g(x)$, one finds the area under the curve of the product $f(x)g(x)$, which was multiplied by six to show its shape.

In the second line the differential $dx \to \Delta x$, and the integral is replaced by a sum. The more primitive expression is a step forward, not backward from calculus. First, and to repeat what we say elsewhere, almost all the integral signs in quantum mechanics stand for numbers, not anti-derivatives, and numbers are easy. Second, the sum in equation (9.4) represents an *inner product* or dot product of lists considered to be vectors on a space of arbitrary dimensions. The inner product is a natural way to compare lists, as well as having so many convenient features it becomes indispensable. When functions are complex the inner product is defined with the complex norm-squared $|z|^2 = |x + \imath y|^2 = x^2 + y^2 = zz^*$, where $z^* = x - \imath y$ is the complex conjugate[1] of z. The overlap is

$$< f|g > = \int_{-\infty}^{\infty} dx\, f^*(x)g(x) \quad \leftarrow \text{overlap of complex } f \text{ and complex } g.$$

Note the * is on the left function by convention: not on both. The order matters. In general $<f|g>^* = <g|f>$ happens to be quite clever, and also needs some attention.

Notice that multiplying a function by a constant, $f(x) \to \lambda f(x)$, can arguably be said not to change its actual 'shape', but just how tall it is. Children would say a picture of Mount Everest and the actual Mountain have the same shape. The norm-squared of $f(x)$ is defined to be $|f|^2 = |<f|f>|^2$. It's just one number equivalent to one overall constant factor. Everything else about a function is in $f(x)/\sqrt{|<f|f>|}$, which is said to be *normalized,* which means its norm is 1. Often without discussion many functions are assumed to be normalized: it is a good organizational trick to 'put the normalization on a shelf' and deal with it separately, when needed.

Older treatments made a major issue of calculating the norms of wave functions the moment they were introduced, and 'implementing the normalization postulate'. As we have seen, the 'postulate' is actually an *agreement* for notation. *We agree to describe* quantum physical systems with normalized wave functions, which (at most) means we agree to not use one number in the overall normalization. (In plain words, if we don't need it, let's not deal with it.) We are reiterating this small point because it saves a lot of work. If your sources spend a lot of effort calculating normalizations, they are actually not explicitly needed, until they are needed.

Back to comparing waves: if the overlap is zero, two functions are *orthogonal,* and quite unlike one another (see figure 9.5). We can say orthogonal functions are *mutually exclusive*: there is no trace of one in the other. Meanwhile if the overlap of two normalized functions is one, they must be the same function. The full range of overlaps of normalized functions is zero (minimum) to one (maximum):

$$0 \leq |<f|g>|^2 \leq 1, \quad \leftarrow \text{full range of overlaps,}$$

repeating that $|f| = |g| = 1$. The (normalized) overlap is only one number, and rather insensitive to the differences of wave shapes, but not completely insensitive.

[1] In case your math is rusty, $z^* = x - \imath y$ is *a definition* concocted so the Pythagorean quantity $z^*z = (x - \imath y)(x + \imath y) \to x^2 + y^2$ comes out right.

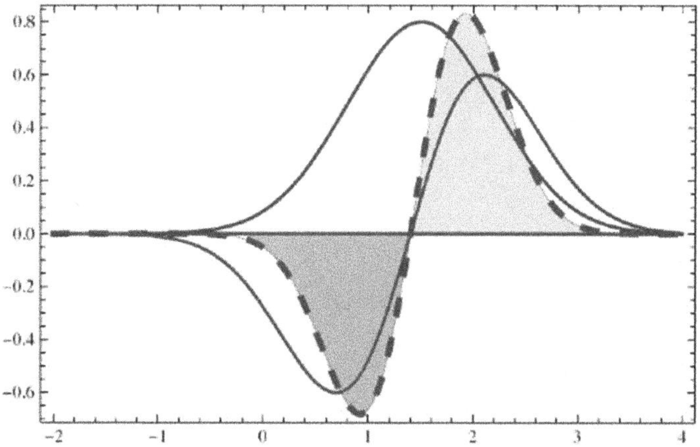

Figure 9.5. Two functions *f*, *g* are completely independent if they are *orthogonal*, with zero overlap. This means they are either completely separated, or that positive and negative regions of $f(x)g(x)$ (shaded) cancel in computing $< f|g >$.

Yet the Universe of shapes is so vast it is *very useful* to have a single-number comparison between two functions.

9.1.6 The Born rule of quantum probability

We come to the *Born rule*, which tells you the quantum mechanical probability that *your description of a system* can use wave function ψ_2, given a wave function ψ_1. We don't think the Born rule is a 'fact of nature', while some do. We claim the Born rule is a fact of our choice of description and exploiting the most convenient mathematical opportunities. It is not derived from probability as 'frequency'. The Born rule essentially begins with a fact *nobody can stop us from choosing to define probability using it*. We then explore the implicit classes that turn out to be distinguished under the rule. That sounds vague, and the Born rule in fact is deliberately vague, for a good purpose.

The Born rule uses the fact that any candidate ψ_2 already pre-exists in ψ_1—by mathematics—with an amplitude one can compute. That component will remain after a non-too-destructive 'filtering' (or projective) process reveals it. For example, a continuous superposition of wavelengths can be separated by a diffraction grating scattering the incoming wave into a range of angles. If the grating is ideal, each Fourier component labeled by $\vec{k}$ coming in will be filtered and separated to emerge in a particular direction $\vec{k}'$. The process is non-destructive and in principle reversible. For example, sending light through glass prisms glued together to make a 'polarizing beam-splitter' will be sorted into perpendicular x and y polarization amplitudes. Another beam-splitter downstream, and tilted at an arbitrary angle, will extract whatever amplitudes are available, and sort them again.

To express the filter effects mathematically, let $|\psi_1>$ be the given wave. The amount of $|\psi_2>$ pre-existing in $|\psi_1>$ is given by the projection. The relation is

$$|\psi_1> = |\psi_2> < \psi_2|\psi_1> + (|\psi_1> - |\psi_2> < \psi_2|\psi_1>). \quad (9.5)$$

Look hard at the equation: it says $|\psi_1> = |\psi_1>$, so it is correct. The term in brackets is orthogonal to $|\psi_2>$:

$$< \psi_2|(|\psi_1> - |\psi_2> < \psi_2|\psi_1>) = < \psi_2|\psi_1> - < \psi_2|\psi_2> < \psi_2|\psi_1> = 0,$$

assuming $<\psi_2|\psi_2> = 1$, or ψ_2 is normalized[2]. The first term in equation (9.5) is the part of the initial wave parallel to $|\psi_2>$, and the other part is orthogonal, hence mutually exclusive of $|\psi_2>$. It could be called $\psi_{2\perp}$: see figure 9.3 once again.

The upshot (and obvious to those skilled in the art), is that *given* any ψ_1, you already have ψ_2 with amplitude $<\psi_2|\psi_1>$.

The Born rule *assigns and invents an ad-hoc probability concept* with the following agreement: the Born rule probability is denoted $P(\psi_2|\psi_1)$, where the line | is read 'given', and has the formula

$$P(\psi_2|\psi_1) = |<\psi_2|\psi_1>|^2. \quad (9.6)$$

Since the maximum value of $|<\psi_2|\psi_1>|^2 \to 1$ if and only if $\psi_1 = \psi_2$, the Born rule probability that a function will be itself is $<\psi|\psi> = 1$. The minimum value of $|<\psi_2|\psi_1>|^2 \to 0$ if and only if ψ_1 and ψ_2 are orthogonal, or mutually exclusive: that corresponds to zero probability.

The overlap-*squared* appearing in the Born rule comes from the physics of *intensity*. The intensity of light and sound waves goes like the amplitude-squared. An intensity meter scanned over the range of angles emerging from a diffraction grating will yield intensities going like $|\psi_k|^2$. This fact is universal for all *linear* wave theories. The Schrödinger theory is linear so it obeys the rule. The Born rule will work when an experiment measures some kind of intensity—there are many variations—which has been efficiently and faithfully filtered out from the initial state. The probability of those measurements is given by $P(\psi_2|\psi_1)$. Since *intensity* is ubiquitous in physics, and probably more inevitable than we appreciate, the ad-hoc character of the Born rule is almost already a *theorem*, not a postulate [1].

In section 9.1.3 we *derived* a tentative feature of probability equivalent to the Born rule. It was hard to escape assuming the eigenvalue postulate. Despite our criticisms, the *ep* is not always wrong, and using probability as frequency is not always wrong. To state this positively, it's common to find experimental situations where the Born rule works without needing to think hard about it. In cases where it's not so obvious it is good to accept the Born rule neutrally as a *procedural agreement for our own symbols*, which happens to be precisely what it is.

[2] There's no cheating with normalization postulates needed here. The Born rule is more complicated to write when functions are not normalized, so we're not doing that.

9.1.7 Avoid bunk about disturbance of measurements

The Born rule is often associated with an 'irreducible destruction of information by the uncontrolled interaction of measurement'. That's false, and harmful to understanding. Our analysis shows no uncontrolled interaction and no measurement has been done. The Born rule assumes the process of projective filtering was gentle, and if there is any significant disturbance or destruction of the components filtered out, the rule cannot predict it.

The bunk usually accompanies slippery presentations of the uncertainty relation and pretended interactions imagined using Planck's constant. In that literature the Born rule was actually suppressed. When those probabilities of point-like *parcles* were introduced instead, there was a perceived need to show how a wave packet could have a wide spread of wave numbers, which a *parcle* does not have. The answer was to pretend an interaction created the spread of *parcle* momentum'. *Before* you learn quantum mechanics you cannot tell the argument is a hoax. *After* you understand quantum mechanics, you will know that an interaction requires a definite calculation of time evolution, to see how the wave function changes. Those calculations were not made. The suggestion they were made by some advanced mentality is a sham. So to repeat, the neglect of any interaction that changes the wave function's projection onto the detected component is clear in the Born rule's formulation, equation (9.5).

It is true that to measure a system your apparatus must somehow interact with it. The tradition then creating an artificial distinction between measuring classical and quantum systems has obsolete and unhelpful elements. One problem is that measuring classical systems is taken as trivial and not worth critical examination, while it merits your attention. If one gives classical mechanics the same degree of literalism as quantum mechanics, many of the peculiar limitations attributed to quantum mechanics reappear. For example, suppose a classical system has a sharp and conserved momentum, as the quantum eigenstate postulate asserts. The classical system must be translationally invariant. Then there's no particular position to measure. Such a system exists classically: it is an infinitely long, infinitely smooth rod traveling along its length at a steady speed. One cannot know where it is. This exposes the assumption one is not supposed to treat classical physics seriously. The fundamental objection to schoolbook prescriptions for measurements is that real experiments come in such variety, and with so many inherent unknowns, that it's quite offensive for theoretical authorities to make rules about them in advance. Clever experimentalists have by now demonstrated a nearly unlimited number of quantum measurement stunts that always confirm one part of quantum mechanics while probably violating an authority figure's opinions someplace else. There are even *interaction-free* measurements, where by measuring nothing and *not interacting* with one channel, information about it is obtained through a different channel[3].

Notice that the words 'probability distribution' do not appear in the Born rule. Physics education tends to have a gap in probability theory (misunderstood as

[3] The undergraduate-level textbook by Blümel has many examples.

counting frequencies) and distribution theory, which is used without adequate definitions or notation. A few examples will convince you that quantum probability defined by the Born rule is something new.

Non-distribution example. *Probability distributions* have a sharp definition. Make a graph of the accumulation of 'stuff' labeled N versus a variable x. The rate of change per x is the distribution dN/dx.

Since distributions are rates, they depend on the rate variable in a particular way. A function $y = y(x)$ will lead to a different amount of stuff per y:

$$\frac{dN(y)}{dy} = \frac{dN(x(y))}{dx}\left|\frac{dx}{dy}\right|.$$

This is the chain rule[4] accounting for how the relative rate of y compared to x appears. The first term $dN(x(y))/dx|$ is the distribution in the x variable, replacing x in terms of y. That's an ordinary variable change of a function. The second term $|dx/dy|$ contradicts the rule for replacing variables in functions. You can propose the words 'a distribution is a function', and the words will be wrong, because distributions have a different rule for changing variables. This simple fact is seldom explained, and that explains the perplexity many find with changing variables.

Any function of position $\psi(x)$ has a corresponding function $\tilde{\psi}(k)$ called its Fourier transform. If you know one you can compute the other, and vice versa. The Fourier transform is precisely the overlap of the original function onto harmonic waves. By the Born rule you may consider $\tilde{\psi}(k)^*\tilde{\psi}(k)$ to be the probability to find each harmonic wave of a given k. The most straightforward interpretation will be found in electric engineering. The actual power (intensity) of a generic wave packet $\psi(x)$ that goes straight into a processor or antenna selecting k will be the corresponding $\tilde{\psi}(k)^*\tilde{\psi}(k)$.

Now *if quantum theory had been defined by distributions*, then from a rate $\tilde{\psi}(k)^*\tilde{\psi}(k) \rightarrow dN/dk$ one could predict the rate $dN/dx \rightarrow \psi(x)^*\psi(x)$ by the quaint old 'chain rule':

$$\frac{dN}{dx} \stackrel{?}{=} \frac{dN}{dk}\left|\frac{dk}{dx}\right| \quad \text{(transforming a classical distribution).}$$

But that fails. The Born rule does not generally obey the rules of distributions when variables are changed. In fact, *given* a claim that $\psi^*\psi(x)$ is a distribution, it is not possible in principle to compute $\tilde{\psi}(k)$ or even $\psi(x)$ itself, because phase information was lost.

This is not often noticed, because the 'normalization postulate' tends to hide its most simple examples. In usual form the postulate says without explanation that $<\psi|\psi> =1$. Given the Born rule we predict $<\psi|\psi> =1$, so there's no issue. Suppose we compute $<\psi|\psi> = \int d^3x\, \psi^*(x)\psi(x)$. Change the units of length, say from cm to

[4] The $|dx/dy|$ in the chain rule is both a convention that *stuff N* will be defined to be increasing, and a reminder of the Jacobian rule for transforming multiple differentials.

inches, which is a *scale transformation* $\vec{x} \to x' = \lambda\vec{x}$. Then $d^3x \to \lambda^3 d^3x$, and we must replace $\psi(x) \to \lambda^{-3}\psi(x'/\lambda)$. That transformation IS consistent with a distribution $dN/d^3x \to dN/d^3x'$, usually taken for granted. However it is done 'by hand': it does not come about automatically using a chain rule.

Since it does not come from distributions, the Born rule defines a new extension of probability related to how we use our own descriptions. The new probability will *sometimes but not always* dovetail with the ordinary use of probability distributions. That's why we are presenting the Born rule not as a 'fact of physics', but as a *procedure* which will sometimes give the same result as experimental measurements, when used appropriately. At this point, an authority somewhere may object, saying that 'there are no exceptions to the Born rule'. That's nonsense, because the entire use of wave functions is a special case, often found inadequate, compared to the more general density matrix theory. The variety and complexity of physical measurements is so vast it is not wise to push any 'doctrine of measurement' further.

Reference

[1] Caticha A 1998 *Phys. Rev.* A **57** 157

Chapter 10

Entanglement

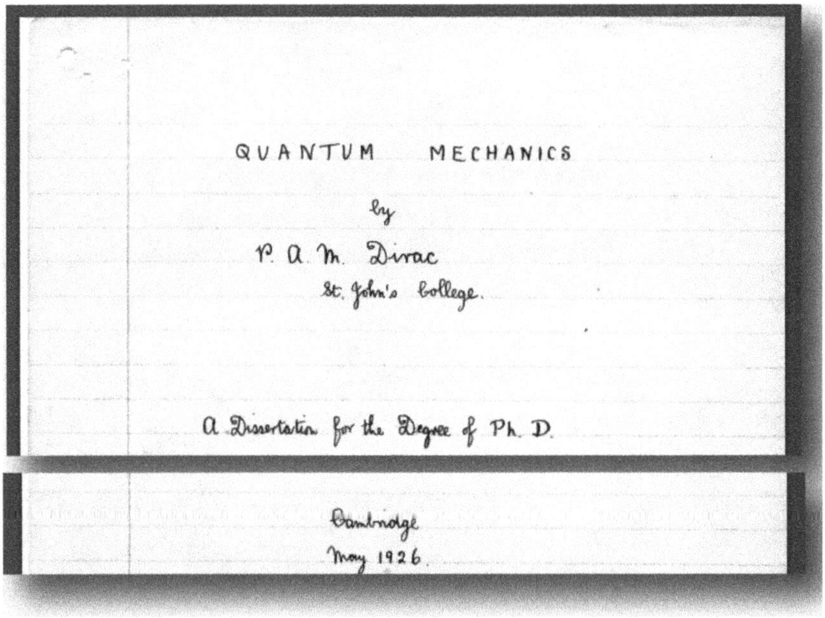

Figure 10.1. The cover sheet from the hand-written dissertation of Dirac, 1926. From the Diginole branch of the Special Collection and Archives of Florida State University.

10.1 Sums of products are generic

In the 19th Century when Maxwell, Boltzmann, Gibbs, and others were developing classical statistical mechanics, the notion of a *classical probability distribution* was rather informal. Probability theory was slow in developing. Decades after quantum mechanics, the great mathematician Kolmogorov realized that mathematicians had forgotten to make axioms for probability. Kolmogorov's superior analysis reduced

doi:10.1088/978-1-6817-4226-7ch10

(his) probability to three axioms for probability P. His axioms say that the probability there is some event someplace in your own sample space is one, the number P is positive and not zero, and there is an additivity rule $P(1 \cup 2) = P(1) + P(2) - P(1 \cap 1)$. While deriving all of *classical* probability was a great feat, Kolmogorov apparently did not notice that his formalism bypassed parts of quantum probability the physicists had been using for years. Physics had jumped out of the axiomatic cookie jar before it was set up.

The Born rule makes a formula for quantum probability without explaining that it is a new mathematical system. The rule transcribes the concept of *mutually exclusive* quantum system to a calculation. Mutually exclusive quantum systems have no possibility of describing one another: they are *orthogonal*, inner product zero. Orthogonal functions are somewhat rare. If your function is a list of 1000 slots, that defines its dimension, and there are exactly 1000 total mutually exclusive normalized functions.

An electrical engineer will classify signals as functions differently. In each slot the signal has N-bits: for eight-bit precision, the signal is resolved into 256 discrete values. Repeated over 1000 slots, there are 256^{1000} mutually exclusive signals. Notice this is much more than 1000! There are only about 10^{80} atoms in the observable Universe. The number of universes in which the engineer needs to store such information is

```
143736034398869180245266113663484224073703334822
433318652402884132868806733604792360415639030443258
300640876089196308150760718393398609942658971312333
583734726940944376523020510445785561419002559483634
718480762416087862406988933546045257558466999076447
574173537634977512700363409437326171391307515795135
492186021853629461747725256008047945133326703447270
417347006244945374658771596390543432238931171001775
014203260161220960695441838028403617456210475768136
116871125723440063971297120833241960985411914470664
023376763842509859419599434165891372665398856944907
990854277200972343489641285536116284082778008910016
277058533820487004495849215640754292438944363029462
464101560372272891531866900312101896936331990300632
663750946939827180604726808665971027800786413200822
625329836670371125829465345020721257320616999081133
270128942582241645638328916449711974012769515472572
460762075205170270887087871161158977210033777999450
379645063265763560048590947604207485166610601430017
513936772253039364727021993731479390837764327893030
909930157645446379807697552437957207792294086543452
897425833343769567879145950767918482990291312234279
953634019214545113770473828674272953002780547290848
311357082860125445397156054861936902354395303300315
```

```
988210253035748525348653267950031343336407250679275
503593877893221626845904050160338130400530238246112
291871170645151518230561913654534217091674199273968
754002885297777167600499657394050747710573065717609
690707514703684272172215882919579627709097613141140
133660882081956870821463770351873920984349507083570
622118241546432430478722774333611760907147295961720
490652470250867891349644039057569248970715887394366
830163142237804004107834750794858496282516381995443
806692934887034031753285370922631143307634427774074
724593672416799433345908958098339063676987447512279
258037824353820088123496865556716478767691296094035
599233781341599988220483483828055741678700878938072
564465556769042880615207994723834454210685879762463
281195922336869754972299546608551494453481250701424
668504614395805474980776262506721909566072078930151
687124024978644946445340719137204186047247542428953
813900070817523215374925121531678311664323950328936
566902359394943131606497266397854590480537701373302
792307511805986762835090489741647765855207141289223
977683174364809046230062470984916437243042806580142
152691032662855824774497835156212233886424436603817
605193236935366036864643399055775761432576/8271806
125530276748714086920699628535658121109008789062
```

Classical probability routinely accepts a classification scheme that is so fabulously detailed it literally cannot possibly *exist*. The particularly vague, unsharp probability classes of the Born rule are ingenious and desirable. The Born rule reduces the effective possibilities under discussion to a relatively small and robust *computational decision*, enormously compressing information. See figure 10.2.

Exactly when will quantum systems be mutually exclusive? We tend to believe quantum systems will be orthogonal when sufficiently separated in space. The wave functions of atoms are somewhat localized, and separation should produce circumstances with zero wave function overlap. Yet this is only tentative, because it is the first time we've discussed more than one quantum system, and interacting quantum systems have new features.

The concept of *two systems* means we can look at one kind of wave at location $\vec{x}_1$, and another wave at location $\vec{x}_2$. A wave function $\psi(\vec{x}_1, \vec{x}_2)$ will be the topic. Any function of two independent variables can be decomposed into a sum of products of functions of each variable:

$$\psi(\vec{x}_1, \vec{x}_2) = \sum_{mn} c_{mn} \upsilon_m(\vec{x}_1) \varphi_n(\vec{x}_1). \quad (10.1)$$

Hence, *products* and features of *multiplication* of subsystems will describe compositions of quantum sub-systems.

A multiplicative feature found in basic probability has often been used to motivate this. When two things have independent probabilities, the probability of both is the product of the separate probabilities

$$P(A, B) = P(A)P(B) \leftarrow \text{classical independence}. \quad (10.2)$$

We cannot use classical probability rules to define the more general features of quantum probability. However, there will be *some cases* where multiplying wave functions plus the Born rule will give the result of independent classical probability, because the theory must cover *all possible cases*.

Once there are products of wave functions, the most common oversimplification puts too much emphasis on *one product*, $\psi_{tot} \rightarrow \psi_1\psi_2$. The hydrogen atom might possibly be described by products of wave functions for the proton and the electron. Let $\vec{x}_e$ be a region where you are considering an electron wave, and $\vec{x}_p$ where considering a proton wave. One might guess $\psi(\vec{x}_e, \vec{x}_p) \overset{?}{\rightarrow} \psi(\vec{x}_e)\psi(\vec{x}_p)$. But a simple product can only represent independent systems, as follows.

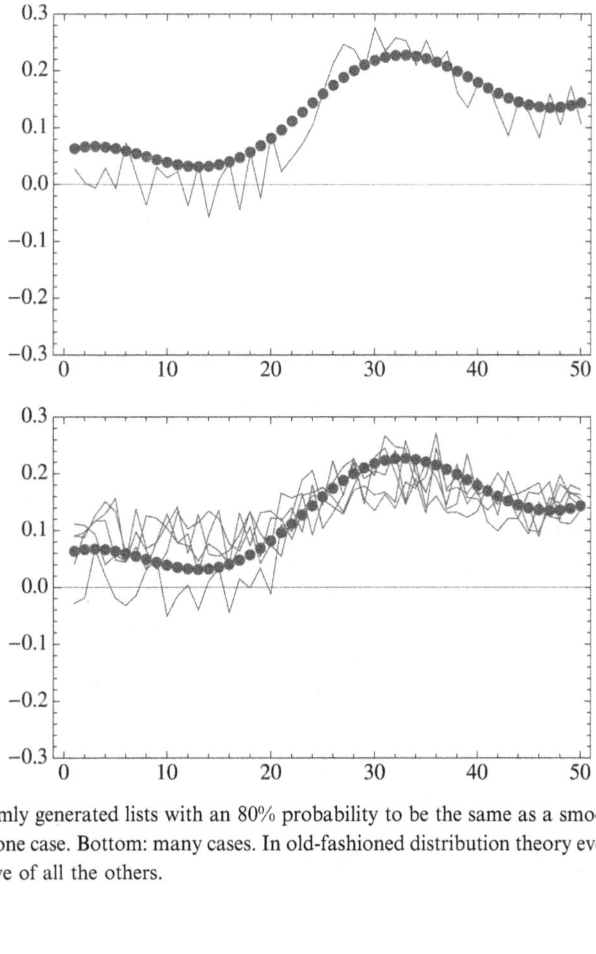

Figure 10.2. Randomly generated lists with an 80% probability to be the same as a smooth list, according to the Born rule. Top: one case. Bottom: many cases. In old-fashioned distribution theory every single case would be mutually exclusive of all the others.

The time evolution will come from the frequency operator. For electron waves there is one operator $\vec{\nabla}_e^2 = \partial^2/\partial \vec{x}_e^2$, and for proton waves $\vec{\nabla}_p^2$. Suppose $\Omega_e \psi_e = \imath \partial \psi_e / \partial t$, with no role for ψ_p. Suppose $\Omega_p \psi_p = \imath \partial \psi_p / \partial t$, with no role for ψ_e. The product obeys

$$\imath \frac{\partial}{\partial t}(\psi_e \psi_p) = \imath \frac{\partial \psi_e}{\partial t} \psi_p + \imath \psi_e \frac{\partial \psi_p}{\partial t} = (\Omega_e + \Omega_p) \psi_e \psi_p.$$

This is an 'addition law for operators' and then an 'addition law for frequencies'. When independent systems are composed, their eigenstate frequencies add. The frequencies of superpositions will include the sums of all terms available.

Now we know what is *not* the outcome of interaction: simple products.

$$\Omega_= \Omega_e + \Omega_p \leftarrow \text{no interaction.}$$

When there is interaction there must be terms that depend on both variables:

$$\Omega_= \Omega_e + \Omega_p + \Omega_{ep} \leftarrow \text{interaction.}$$

The arrow indicates the *interaction term* that couples equations together, and generally makes things difficult to solve. When there are interactions, the generic sums-of-products expansion (equation (10.1)) can't be avoided. And that is a good thing, because existence needs it.

In the 1930s Schrödinger noticed a pattern of probability concept errors was creeping into quantum mechanics, especially from extrapolating classical distribution theory, point-like *parcles*, and so on. To focus attention on the situation of a generic mixed up sum of products, Schrödinger gave it a new name: *entanglement.* He made calculations and examples showing how entanglement caused effects inconsistent with the dumbed-down replacement of electron waves by point-like electron probability clouds. The famous Schrödinger cat paper was written to expose the absurdity and technical errors of the dumbed-down stuff. (The enemies of physics[1] turned it around and claimed Schrödinger's cat was another proof quantum mechanics was not supposed to be understood.) After being around for about 75 years, entanglement is hot in the 21st Century.

Quantum gurus of *entanglement* will tell you it is a uniquely new and weird feature of quantum mechanics that the macroscopic human mind is not supposed to understand, and so on. Yet the description of a tropical storm off Florida cannot be reduced to a single product of a function of the latitude times a function of the longitude. Entanglement is first an ordinary math fact of describing more than one *dimension*, and then an ordinary math fact for describing more than one *system.* Entanglement is also used with the Born rule and quantum probability. The composition of entanglement and the Born rule is what transports physics outside

[1] The enemies do exist, and they are evil. They glory in Schrödinger's cat, without telling you the cat died in mid-1930s Germany by cyanide gas, just as millions were being rounded up for the same fate. It was Schrödinger's bad luck to predict it, and evil to cover it up.

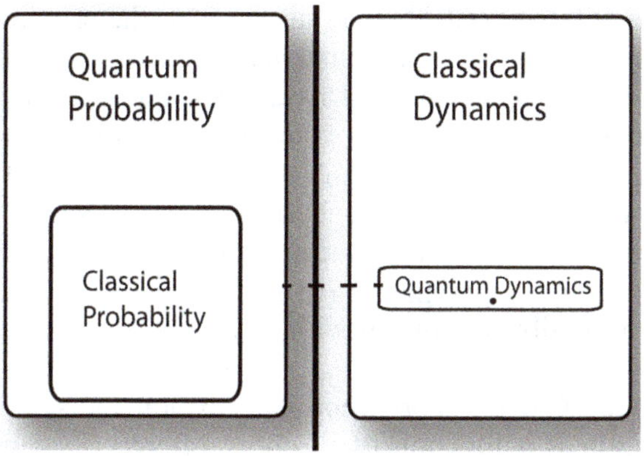

Figure 10.3. The big picture of quantum mechanics repeated. The little dot represents the relation of quantum dynamics to classical Hamiltonian and Lagrangian dynamics.

Mr Kolmogorov's cookie jar. People will often say it proves 'there is no quantum reality'. They are thinking of their mistaken point–*parcle* reality.

10.1.1 The two wave electron–proton atom

What really is an atom? Since it's part of nature, we don't really know. However an atom is described by a *correlation* of electron–proton waves. We cannot predict where atoms will be found in the Universe. Those facts are mostly in the initial conditions. When you find an electron wave in a hydrogen atom, it always comes with a proton wave, and the wave function describes the correlation. Correlation is a very general idea, and not a particular feature of quantum mechanics; see figure 10.4. However quantum theory manages the information of correlations in new ways naturally expressed with wave functions.

We mentioned that entry-level quantum mechanics takes for granted an unphysical fixed 'interaction potential', which in reality should be a dynamical interaction. The

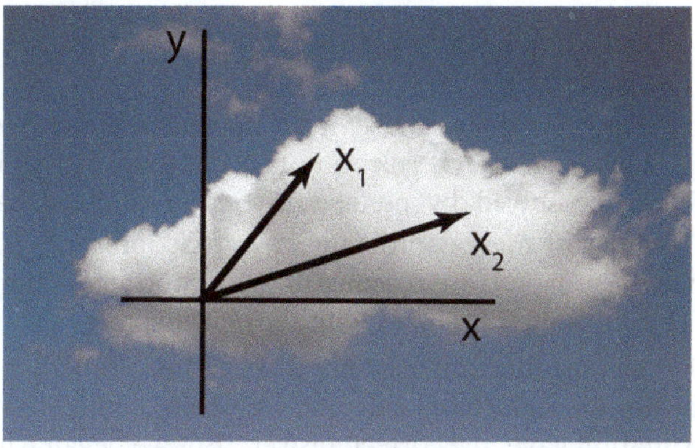

Figure 10.4. It takes two or more position vectors $\vec{x}_1$, $\vec{x}_2$ to describe a correlation. An ordinary cloud of water is described by a correlation of humidity at $\vec{x}_2$ related to humidity at $\vec{x}_1$. Correlations themselves are not a particular discovery of quantum theory. Photo by Michael Jastremski.

mythology of the fixed potential is a central player in the classical Kepler problem, where the Sun is assumed to have an infinite mass, making recoil negligible.

Those flaws are not part of classical mechanics, and they are corrected by the elegant center of mass (CM) and relative coordinate decomposition. It comes from a variable change[2] from two position vectors $\vec{x}_1$, $\vec{x}_2$ to $\vec{r} = \vec{x}_1 - \vec{x}_2$ and $\vec{R} = (m_1\vec{x}_1 + m_2\vec{x}_2)/(m_1 + m_2)$. In Newtonian physics m_i are the Newtonian masses, while for quantum waves $m_i \to \mu_Q$ are the quantum mass parameters: the math works the same. We never refer to Newtonian mass, so we can use symbol $\mu_e \to m_e$ in this section, to avoid a possible confusion with a reduced mass symbol μ that is very common.

The reason for the CM transformation is to exploit the symmetry of problems where the interaction is a function of position differences, namely $\vec{r}$, and not a function of the position of the origin, carried in $\vec{R}$. Due to the *symmetry* that $\vec{R}$ has no preferred location, its dynamics is trivial in both the Newtonian and quantum wave cases, effectively reducing the dimension of the problem.

We are ready to describe the joint electron–proton system. It will involve a product-space wave function $\psi(\vec{x}_e, \vec{x}_p)$ expressing information at two independently selected points. The conventional Schrödinger Hamiltonian operator will be $H = -\vec{\nabla}_e^2/2m_e - \vec{\nabla}_p^2/2m_P + V(\vec{r})$. After the change of variables

$$H = \frac{-\vec{\nabla}_R^2}{2(m_e + m_P)} - \frac{(m_e + m_P)\vec{\nabla}_r^2}{2m_e m_P} + V(\vec{r}) = \Omega(R) + \Omega(r). \tag{10.3}$$

Symbol $\vec{\nabla}_R^2 = \partial^2/\partial\vec{R}^2$, and so on. As always $V(\vec{r})$ is an interaction function. Using $V = -\alpha c/r$ predicts the *physical* hydrogen atom and its *correlation* in terms of $\vec{r} = \vec{x}_e - \vec{x}_p$, and the frequency parameter $m_e m_P/(m_e + m_P)$. Already this is new information. For deuterium with nuclear mass (frequency) $m_d \sim 2.01 m_P$, replace $m_P \to m_d$, which will cause a change in the predicted spectral frequencies.

The frequency operator is a sum of $\Omega(R) + \Omega(r)$, suggesting its eigenfunctions will be simple products. For hydrogen the result is

$$\Phi_{n\ell\vec{P}}(\vec{x}_e, \vec{x}_p) = e^{i\vec{P}\cdot\vec{R}}\phi_{n\ell}(\vec{r}). \tag{10.4}$$

Here $\phi_{n\ell}$ are the eigenfunctions solved as if there were no dynamical proton (equation (7.19)). This is a general rule: any two quantum waves interacting with a function of $\vec{r}$ can be solved by pretending there is a fixed background, and multiplying by a center of mass coordinate wave function $e^{i\vec{P}\cdot\vec{R}}$. If you ignore the internal coordinates $\phi_{n\ell}$, the quantum center of mass coordinate 'thinks' it is moving in free space. Yet don't forget it comes with the $\phi_{n\ell}$ functions.

Equation (10.4) has the function arguments $(\vec{x}_e, \vec{x}_p)$ on the left, and $(\vec{r}, \vec{R})$ on the right. One might use the right hand side and move on, complacent that the fixed-potential approximation gave 'the correct answer'. But we began with interest in the

[2] In this section the symbol $\vec{r}$ is not an alternative for $\vec{x}$, but instead $\vec{r} = \vec{x}_1 - \vec{x}_2$.

electron wave and proton wave examined at the points $\vec{x}_e$, $\vec{x}_p$ on the left side. In terms of those variables

$$\Phi_{n\ell\vec{P}}(\vec{x}_e, \vec{x}_p) = e^{\iota\vec{P}\cdot(m_e\vec{x}_e + m_P\vec{x}_p)/(m_e+m_P)}\phi^*_{n\ell m}(\vec{x}_p - \vec{x}_e).$$

This is distressingly intricate. When solved for a fixed background, the eigenfunctions ϕ_n described the electron wave. Now the coordinate $\vec{x}_e$ appears in two places. The bound state wave function does *not* refer to the electron, but describes a correlation of differences of position where you look at the electron and proton waves: it is a function of $\vec{x}_p - \vec{x}_e$, not $\vec{x}_e$! The hydrogen spectrum comes from oscillations of those *relative correlations,* which were incorrectly[3] identified as purely electron wave oscillations before. See figure 10.4, which reminds you that a cloud-cloud (not an electron cloud) is a *correlation* of water drops probed at many points.

The time dependence is also surprisingly complicated. The frequency eigenvalues associated with the *cm* coordinate are $\omega(\vec{P}) = \vec{P}^2/(2(m_e + m_P))$. Including the internal atomic frequencies, the time evolution is

$$\Phi_{n\ell\vec{P}}(\vec{x}_e, \vec{x}_p, t) = e^{-\iota\vec{P}^2/(2(m_e+m_P))}e^{\iota\omega_H/n^2}\Phi_{n\ell\vec{P}}(\vec{x}_e, \vec{x}_p, t = 0).$$

The formula predicts that the electron's electric current depends on the overall motion of the atom. That ultimately will lead to a Doppler shift of atomic spectra, among other things.

What should we generally do with the $\vec{P}$ dependence? It describes initial conditions. We can imagine a quantum hydrogen atom moving with a sharply fixed $\vec{P}$ and coming right at us. Where is the electron? If you observe nothing about the proton labels, they will be integrated over in observables. The average position of the electron wave will be

$$<\vec{x}_e> = \int d^3x_p < \Psi(\vec{x}_e, \vec{x}_p)|\vec{x}_e|\Psi(\vec{x}_e, \vec{x}_p) > .$$

We suppressed the indices of Ψ. By a neat accident, the plane wave $e^{\iota\vec{P}\cdot\vec{R}}$ cancels out, and the answer tries to be just the same as using a fixed potential. But the relative coordinate is not $\vec{x}_e$ but $\vec{x}_p - \vec{x}_e$. Integrating over $\vec{x}_p$ averages over every point in space: the result of observing the electron's position on its own returns no position! Actually that makes sense: the electron and proton come together, and the atom is a *correlation* of the two. We should have looked at observables depending on $\vec{x}_e - \vec{x}_p$. They correspond to what's done 'ignoring' our discussion and using a fixed background interaction.

The most simple case would be to choose $\vec{P} = 0$, 'an atom at rest'. That is not realistic: the function $e^{\iota 0\times\vec{R}}$ is constant in $\vec{R}$, and extends *everywhere over space.* We generally believe the atom is not smeared over everywhere, but should have $\vec{R}$ variables localized within a small region. That will be represented by superpositions

[3] The identification was correct for the model used, and avoided complicating things too early.

of plane waves $\int d^3 P e^{\imath \vec{P} \cdot \vec{R}}$..., which predicts the correlations of $\vec{x}_e$ and $\vec{x}_p$. There are few or no cases where we can conceptually replace the difference $\vec{x}_e - \vec{x}_p \to \vec{x}_e$ and not make a mistake.

Entanglement has shocking lessons about how quantum physics works. The number of possible *entangled relationships* in a simple physical system can be huge. When different situations and concepts are identified one by one, physics is usually smarter than we are, because all the different combinations are already there waiting to be noticed!

10.2 Promoting operators, and other notation issues

As we mentioned, entanglement starts with the basic fact that generic functions of two variables x, y are sums of all possible functions of each variable. Once the operator $\partial/\partial x$ is defined for functions of x, an agreement is needed for what the operator does to functions of y. The agreement is that $(\partial/\partial x)g(y)f(x)= g(y)(\partial/\partial x)f(x)$. This is not actually obvious from $\partial y/\partial x = 0$, which might mean that $(\partial/\partial x)$ acts like '0' on $g(y)$. The actual agreement is that $(\partial/\partial x)$ acts like '1' on $g(y)$. Then $\partial y/\partial x = y\partial 1/\partial x = 0$ also follows. At one point in history (and often in thermodynamics) a more complete notation $\partial/\partial x|_y$ made the meaning explicit.

A convention in quantum mechanics agrees that an operator on variables a is promoted to act like 1 on *independent* variables b, unless something different is specified. The convention makes a convenient, but also sloppy notation. Sloppy notation with products, operators, and entanglement can cause experts to get confused, so you should be aware of it like a dangerous snake.

Next, we look at the $|\psi\rangle$ and $\langle\psi|\psi\rangle$ notation attributed to Dirac, but really just adopted from mathematicians' (ψ, ψ) inner products. There is a conceptual reason for the $|\psi\rangle$ notation, which seeks to represent objects in 'coordinate-free' fashion. Nature does not depend on our basis, so that classy notation seeks to avoid using a basis. That's why the symbol $|\psi\rangle$ stands for the state as a thing in itself, and separate from the basis used to express components. However this has a potential cost of omitting the information needed to be specific. In the basis of space coordinate x, each slot x represents a mutually exclusive unit vector called $|x\rangle$. The notation ψ_x stands for a list indexed by x: the precise relation is

$$\psi_x = \langle x|\psi\rangle .$$

Since it was not needed at the time, we deliberately bypassed this nuance in our early use of $|\psi\rangle$ notation. Also, the left side uses two keystrokes, the right side uses five keystrokes, and even more pencil motions, so the right side notation is not always superior. Entanglement has upgraded the math level, and wants a coordinate-free notation, but as mentioned, When the Dirac notation is extended to symbols like $|a\rangle|b\rangle|c\rangle$, the notation tends to omit too many symbols to reliably specify what is intended. Adept experts use their experience to make the notation work, but even experts can be confused. Soon enough,

Dirac notation evolved to expressions like $|\vec{k}_1, \vec{k}_2, \vec{k}_2 ... >$, which represents the direct product of three plane waves with labels $\vec{k}_i$ and the spatial coordinates $\vec{x}_i$ entirely omitted. Then

$$<\vec{x}_1, \vec{x}_2, \vec{x}_3 ... | \vec{k}_1, \vec{k}_2, \vec{k}_3 ... > = e^{i\vec{k}_1 \cdot \vec{x}_1} e^{i\vec{k}_2 \cdot \vec{x}_2} e^{i\vec{k}_3 \cdot \vec{x}_3} ....$$

Each different notation has a commendable function in some context where it is most convenient. Simply knowing that *the notation itself can be a barrier* is a major part of understanding entanglement.

Be warned that careless writers will inadvertently give false information originating in virtuoso calculations (sometimes) presented with sloppy word abuse and sloppy notation. It is very common to hear that 'entry-level quantum mechanics is based on wave functions, but quantum field theory is based on operators'. That is because manipulating certain operators can sometimes give the same outcome as computing with wave functions. If you pursue this, you will discover that quantum wave functions like $|\vec{k}_1, \vec{k}_2, \vec{k}_2 ... >$ are still present under the smokescreen of operator-based computational technology. It is *not* true that the Schrödinger wave function became obsolete and was superseded by later developments. We also don't see why the wave function would become 'unreal' for having any number of indices. It may be a proxy for reality, but its dynamics act as real as anything else is real.

10.2.1 The quantum interferometer

Dirac's 1930 textbook—*Principles of Quantum Mechanics*—was a masterpiece, despite the fact Dirac was only 28 years old. Dirac had gone into quantum physics early. He recognized Bohr's influence as a philosopher, but wanted ideas that could be expressed with equations. While making his own contributions Dirac acted as a formalizer and an arbitrator of the turf wars between Heisenberg, Born, Bohr, and Schrödinger.

The brilliant introduction of Dirac's book sets up a paradox. Consider an interferometer (figure 10.5), which divides a ray of light into two beams traveling two paths, which evolve with phase shifts and are recombined. The interference pattern observed after recombination proves light is a wave. Operate the device at very low intensity so you can count photons. Reduce the intensity so two photons

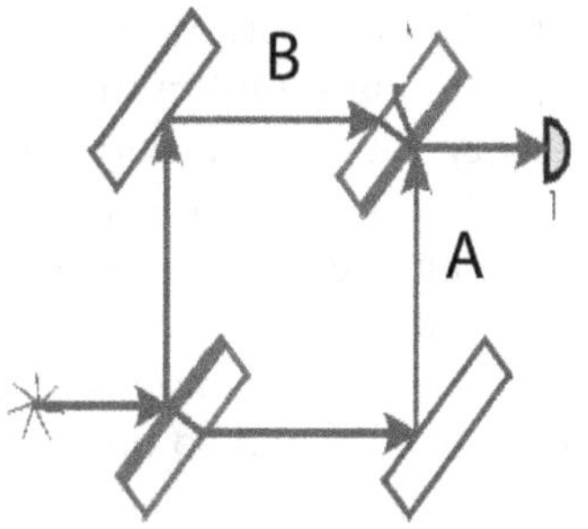

Figure 10.5. A cartoon of an interferometer. It's usually forgotten that 'nothing' on path *A* or *B* is the 'something to specify' called the ground state.

cannot be present at any one time. (The experiment was done early[4] with filters to reduce the intensity, and waiting months for the interference pattern to be recorded on photographic film.) So long as two paths are open, the interference pattern is obtained. This is *prima facie* evidence that describing quantum systems with waves is needed, even when 'counting photons'. The behavior of orthogonal light polarizations, which do not interfere, is also just as classical wave physics predicts. Yet putting a photon detector ('phototube') in either arm of the interferometer produces one 'pop' of the instrument at a time, recording one whole photon. A whole photon is that thing giving a transition resonantly matching frequencies $\omega_\gamma = \omega_1 - \omega_2$, which if not a general rule of nature, is what photons and atoms do under ideal photon-detector conditions. No half photons are observed. No Newtonian dodge that one particle split into two, and joined up later, can be made to work.

In an unusually long (for Dirac) discussion, Dirac concluded in his own words that both and neither waves and particles must be a necessary interpretation of quantum mechanics. Formalizing this compromise dictated by Bohr kept the *parcle* concept in play. Another compromise was giving indeterminacy early development, because Dirac had finessed the steps to make it mathematically consistent. However Dirac was unwilling to write equations solely for the purpose of producing contradictions. Under the next topic *Superposition and indeterminacy*, Dirac wrote apologetically

> 'The reader may feel dissatisfied .... He may argue that a very strange idea has been introduced—the probability of a photon being partly in each of two states of polarization, or partly in each of two separate beams .... He may say further that this strange idea did not provide any (new) information about the experimental result... it may be remarked that the main object of physical science is not the provision of pictures, it is the formulation of laws....'

Here Dirac expressed his preference for equations, not pictures. Unfortunately, Dirac had not given any equations for the interferometer: it's clear had had none to offer. It was too early for him to be in command of entanglement. As a result of the interferometer being botched, mixups started happening.

The interferometer can be understood using the appropriate *entangled* products. Let $|1_A>$ represent a photon in interferometer arm A. Don't think the symbol implies a profound thought, nor an exhaustive description: quantum physics exploits under-description. Let $|0_A>$ represent nothing found in interferometer arm A. Symbol $|0>$ appears often in quantum theory for 'the ground state', 'the vacuum', etc. Once again, a one year course in quantum field theory can give you an infinitely detained representation of $|0>$, which being the ground state is really interesting, but the symbol under-describes it. You can actually measure $|0_A>$ OR another state $|0_B>$ if you put a detector in arm A OR arm B and find nothing with high probability. It is the same with measuring a photon $|1_A>$ and so on.

[4] Kroemer's textbook cites [1]. This book also mentions the density matrix with a single phrase that it 'would go beyond the scope of this text'.

Consider products of these symbols. The product $|0_A> |0_B>$ describes instruments registering nothing in A AND strictly correlated with nothing in B. A strict correlation of two nothings is expected when the illumination of the interferometer has been turned off. (It is a small accomplishment to translate the experimental situation to symbols.) Consider the product $|0_A> |1_B>$. It describes finding nothing in A AND finding one photon in B. That describes half the data. The product $|1_A> |0_B>$ describes the cases of finding one photon in A and nothing in B. This describes the other half of the data. These two cases are mutually exclusive, and orthogonal, by construction[5]. To describe equal probabilities of 1/2 under the Born rule, we propose an interferometer *state* $|INT>$:

$$|INT> = \frac{1}{\sqrt{2}}|0_A> |1_B> + \frac{1}{\sqrt{2}}|1_A> |0_B> .$$

This was not written down in Dirac's book, and seems to be hard to find written anywhere. It is also incomplete. After evolving a distance x_A (x_B) on side A (B) we propose $|1_{A(B)}> \to e^{-ik_{A(B)}x_{A(B)}}|1_{A(B)}>$. Finally there will be a rule for recombining beams, and a rule to read out the final intensity.

It is interesting to ask why such a simple thing as an interferometer has such a devious description. It is partly because we are not accustomed to working with products. We are also over-trained to ignore $|0>$: the vacuum is not nothingness, it is the ground state, and it exists. The interferometer is like a musical instrument. It has two resonant arms coupled by the half-silvered mirror at the front end. The innocuous half-silvered mirror excites both the $|0_A> |1_B>$ and $|1_A> |0_B>$ vibrational modes at the same time: it is not unlike one jet of air exciting two organ pipes of the quantum electromagnetic field.

Given these secrets you can invent and explore non-standard interferometers. Consider $|INT_2> = |1_A> |0_B> /\sqrt{2} + |1_A> |1_B> /\sqrt{2}$. The symbols predict 50% of events[6] with one photon in A and nothing in B, and 50% of events with one photon in A AND one photon in B, which we never hear about. Is that possible? Why not! An enemy letting light into side B might make it happen. Note that $|INT_2>$ is orthogonal to a state $|INT_{2\perp}> = |1_A> |0_B> /\sqrt{2} - |1_A> |1_B> /\sqrt{2}$:

$$\begin{aligned} < INT_{2\perp}|INT_2> &= \left(\frac{1}{\sqrt{2}} < 1_A| < 0_B| - \frac{1}{\sqrt{2}} < 1_A| < 1_B|\right) \\ &\quad \left(\frac{1}{\sqrt{2}}|1_A> |0_B> + \frac{1}{\sqrt{2}}|1_A> |1_B>\right) \\ &= 0 + 0 + (\frac{1}{2} < 1_A| < 0_B||1_A> |0_B> - \frac{1}{2} < 1_A| < 1_B||1_A> |1_B>) \\ &= \frac{1}{2} - \frac{1}{2} = 0. \end{aligned}$$

The algebra is very ordinary, but if you find the notation is *awful*, you are right! Anyway, your theory of $|INT_2>$ is predicting no cases where $|INT_{2\perp}>$ will serve. Was that your intention?

[5] $<0_a|0_b> = <1_a|1_b> = \delta_{ab}$.

[6] We're using probability as frequency, which is not always wrong.

Consider $|INT_3> = |1_A> |0_B> /\sqrt{2} + |0_A> (|1_B> + |2_B>)/\sqrt{4}$. Observing nothing on side A is strictly correlated with seeing one photon in B half the time, and two photons the other half of the time. With the same conditions, side B will also respond 100% of the time to an instrument detecting $|1_B> + |2_B>)/\sqrt{2}$. Can this exist? Why not!

We are learning here that *quantum mechanics is descriptive, not restrictive.* It is rarely *predictive*, because each experimental situation tends to need detailed study to know what it 'is'. Quantum mechanics, even including the extreme version of the eigenvalue postulate, does not actually contain the restrictions on measurements assumed in the *OQT*. Notation should also not be viewed as making predictions or restrictions on reality. Instead the symbols represent possible systems you might find if you actually study reality. *This is a breakthrough in understanding quantum systems.* The bookkeeping is so flexible it can describe many things you never imagined. *What actually occurs* in nature is a separate issue, which no amount of notation will decide!

Continuing, the big discovery of entanglement is that rather simple sums of products of wave functions describe *correlations* of outcomes with a variety never before considered. If your picture of reality is a point-like localized photon it will never explain those correlations: so give up that picture. From the start we've said the wave function is not nature itself, but a proxy used to describe nature. There is also *not enough information in the description* $|0_A> |1_B> + |1_A> |0_B>$ to know whether each individual event is being described. If the description looks good after 1000 events, then it describes 1000 events. There are so many alternative possibilities that asking for one cookie-cutter to describe every individual case cannot be imposed from the top down. That is why Einstein wrote: 'The attempt to conceive the quantum-theoretical description as the complete description of the individual systems leads to unnatural theoretical interpretations', which one can easily avoid. Unless you have some advance knowledge of which wave functions are 'allowed', how will you know which entangled wave functions are disallowed?

10.3 The Stern–Gerlach experiment

The *Stern–Gerlach* experiment (1922) provides another example of entanglement. If you understand entanglement you understand the experiment, and vice versa.

The topic is also botched very often. *Wikipedia* begins:

'The Stern–Gerlach experiment demonstrated that the spatial orientation of angular momentum is quantized.'

No it did not. The usual story says an electron beam goes through a magnetic field, and out come two mysterious beams, labeled by their spins 'up' or 'down'[7]. This (goes the story) proves there is an inherently quantized measurement of 'spin-up, spin-down', which also 'cannot be explained or understood', since point particles cannot spin. The experiment will often be presented in the first chapter of a typical book[8]

[7] Actually Stern and Gerlach used a beam of atomic silver, and the experiment won't work with electrons. The quantum postulates old presentations seek to validate won't prevent unphysical mistakes.

[8] Sakurai.

before the concept of a wave function or equation of motion exists. Such presentations encourage the misconception that quantization is inherent in 'spin' and everything else, and that electrons are always in one spin state or another. Perhaps 200 pages later in the same book, spin will actually be defined, but usually as an isolated topic *again divorced* from any wave equation.

The presentation recapitulates another mistaken path of physics history, and makes it appear mandatory. Early physicists—Einstein and Bohr especially—grievously misinterpreted the Stern–Gerlach experiment as a feature of point *parcles*, overlaid with yet another *OQT* quantization principle that 'could not be explained'. The effort to keep the old non-explanation coexisting with quantum mechanics which *does* explain everything causes great confusion.

Here's how it actually works. There exist quantum *waves with polarization.* Electron waves (silver atom waves, etc) happen to have a polarization degree of freedom as valid and concrete as the polarization for light. *Given the polarized wave equation,* the solutions predict the wave propagating in a magnetic field splits into two waves. Two waves separate by dynamics and time evolution. No magic is involved. *Without a polarized wave equation,* the experiment appears to be impossible to understand in the pre-quantum framework of *parcles* assumed. That is true about the wrong framework. Needing a new axiom about the Universe—an eigenstate-type postulate—was true in the mistaken context, and not true in the actual theory.

The mysterious quantization of the Stern–Gerlach experiment comes from a math fact. Any two-way entangled function (a polarized wave) depending on position $\vec{x}$ and polarization $\vec{\varepsilon}$ can be decomposed into sums of products of functions. You can call this college math, or you can call it entanglement. In terms of indices xj for position x and polarization component j, the theorem is

$$\psi_{xj} = \sum_{\alpha}^{All} \varepsilon_j^{(\alpha)} \Lambda^{(\alpha)} \phi_x^{(\alpha)}$$

$$\text{where} \quad <\varepsilon^{(\alpha)}|\varepsilon^{(\beta)}> = \delta^{\alpha\beta}; \quad <\phi(\alpha)|\phi(\beta)> = \delta^{\alpha\beta}.$$

The number of terms *All* ranges up to the dimension of the smaller space. This is remarkable: if $All = 2$ (or 3, or 5, …) total polarizations labeled j, it is very small compared to the dimension of the space labeled by x. The theorem predicts a decomposition exists that is much more simple than all possible products found in a general expansion in a generic basis. The theorem was known[9] from about 1900 as the *Schmidt decomposition*, but apparently not correctly used (if known at all) by physicists when quantum mechanics was in formation. The theorem says *you can always find two special bases* for the space and polarization variables which are diagonal, as shown. The constants $\Lambda^{(\alpha)}$ are relative weights. Writing out two terms with $\Lambda^{(\alpha)} = 1$ gives

$$\vec{\psi}(\vec{x}) = \vec{\varepsilon}^{(1)}\phi^{(1)}(\vec{x}) + \vec{\varepsilon}^{(2)}\phi^{(2)}(\vec{x}).$$

[9] It is also called the 'singular value decomposition' and often included in modern software packages.

The formula represents a *strict correlation* that polarization $\vec{\varepsilon}^{(1)}$ is found with spatial configuration $\phi^{(1)}(x)$, and ditto for $\phi^{(2)}(x)$. Not only is the decomposition most simple, but also the factors ε_j and ϕ_j are each orthogonal on their own spaces. When the beams $\phi^{(1)}$ and $\phi^{(2)}$ are spatially well-separated as the experiment arranged, they are surely orthogonal. Then when you find a beam at position 1, it is *strictly and always correlated* with polarization $\vec{\varepsilon}^{(1)}$, which is mutually exclusive of $\vec{\varepsilon}^{(2)}$ found in the beam at position 2. This is mathematical magic (but not magic) those insisting on a separate eigenvalue postulate evidently did not recognize. (Hence the reversion to Copenhagen habits of not explaining by postulating what was observed.) Yet mathematical theorems don't count as discoveries about the Universe: all mathematical facts were present from the start in your mode of description, whether or not it took years to recognize it.

Here is an embarrassing fact. Apparently Einstein and Bohr in 1922 were unaware of a related Stern–Gerlach experiment done before 1830. We have no record they ever considered it. Light rays passing through a good calcite (Iceland spar) crystal split into two beams. Each beam has a definite polarization; see figures 10.6 and 10.7. The mechanism is basic wave mechanics. When a particular polarization has a distinct propagation speed, Snell's law (a wave fact) tells you the rays will separate. Malus, Fresnel, Arago, and many others had this completely in hand

§ 4. *Double Refraction of Light explained by the Wave Theory.*

By means of Iceland spar cut in the proper direction, double refraction is capable of easy illustration. Causing the beam which builds the image of our carbon-points to pass through the spar, the single image is instantly

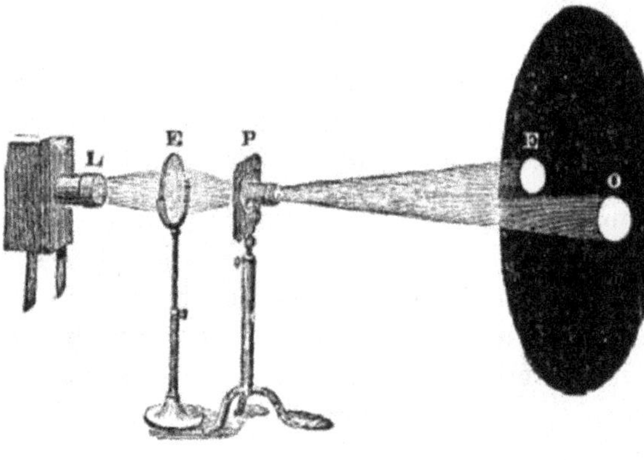

Fig. 26.

divided into two. Projecting (by the lens E, fig. 26) an image of the aperture (L) through which the light issues from the electric lamp, and introducing the spar (P), two luminous disks (E O) appear immediately upon the screen instead of one.

Figure 10.6. The Stern–Gerlach experiment discovered *double refraction* of matter waves. It had long been known for light. The figure is from Tyndall J 1875 *Six Lectures on Light Delivered in the United States in 1872–1873* (London: Longman).

Figure 10.7. A single small light source is behind the calcite crystal in the author's hand. The beam of light splits into two beams, much as in the Stern–Gerlach experiment, supposedly requiring a new quantum principle. However the splitting of light beams known as 'double refraction' was completely understood from wave physics before 1830. It was a major clue to the polarization of matter waves completely overlooked by the 'founders of quantum mechanics'.

and explained by 1830. Einstein and Bohr heard from Stern and Gerlach about two spots, but were so focused on *parcles* they never considered the wave mechanism of *polarization*. The lapse is hard to explain. Wave equations and polarization of the electromagnetic field were the *established* physics at the time.

The connection between Stern–Gerlach and polarization is not a great leap. When the writer, who had played with double refraction by calcite crystals as a boy, first heard about the Stern–Gerlach spots as a student, he asked whether the calcite spots were the same thing. The most competent physicist available to consult, Professor Philip Altick, was not entirely sure. He first said the calcite experiment would need to be done with photons. But if you can see light with your eyes, are high-tech photon detectors going to disagree? And Stern–Gerlach measured spots of silver plated on a paper card, not atoms one at a time[10]. (That is, they did *not*

[10] Since individual eigenstates were *not* observed by Stern and Gerlach, we have proposed the experiment measured the *spot operator*, whose eigenstates are eigenspots. They were invisible (not observed!) until developed by the *cigar smoke operator*, used to blacken them.

measure eigenstates of 'spins', up or down!) Yet if we can understand and explain calcite with ordinary wave physics, WHY IS NEW QUANTUM PRINCIPLE OF THE EIGENSTATE POSTULATE MANDATORY AT ALL? Professor Altick was careful, thoughtful, but not entirely sure. He had asked his professors similar questions, and they told him such questions were considered improper, off-base, ignorant. Their general wisdom was that the Principles of Quantum Theory were so profound and sublime that they would invariably predict some things that were previously understood on a physical basis. The *first principle* from Bohr was that quantum mechanics could not be understood. The *consequences* of the first principle were that you were ignorant to be concerned with things that *could be understood* (!). Dirac's first chapter makes nearly the same observation dismissing *polarization*. We suggest you read it.

10.3.1 The relation of polarization and spin

Long before Stern and Gerlach, a certain two-ness of electrons had been suspected from the Zeeman effect that showed spectral lines were doubled in magnetic fields. (Doubling means the fundamental electron frequencies come in pairs: the difference frequencies involve all allowed combinations.) Pauli was famous for suggesting an abstract two-ness exclusion principle before quantum mechanics was discovered. There is nothing magic about 'two' and emphasis on 'the two-state spin' of the electron wave generates many conceptual errors. It is *very important* to know that quantum wave systems exist with polarizations in 2, 3, 4, 5, … integer dimension classes. The Stern–Gerlach experiment might have been done with deuterium, whose nuclear wave has three polarizations, making an ordinary 3-vector wave. That would give three spots. (Actually, the electron in a deuterium atom is more magnetically reactive than the nucleus. There might be six spots for 3×2 total polarization possibilities.)

The first mention of polarization in quantum theory we found comes from C G Darwin [2] in 1927, who suggested 3-vector polarization, avoiding the two-component case which he thought was taboo:

> 'When what is required is to double the number of states of the electron… the wave mechanics (in this the matrix mechanics is better) definitely excludes half quantum numbers for the spin, and so would lead to triplets -1, 0, −1 instead of doublets 1/2, −1/2, … wave mechanics must have suggested to many a way out of these difficulties by assimilating the electron to a transverse rather than a longitudinal wave, for this at once provides the number of states with the necessary factor 2.'

We don't know who wrongly told Darwin the 'doublet' of two states (half quantum numbers) was 'definitely excluded'. Darwin seems to know how tricky light is to have exactly two transverse modes and *three* independent polarizations. Instead of recycling the electromagnetic wave equation (4.2), Darwin explored a very elaborate theory that failed. That left Pauli to propose the two-component Pauli equation for the spin-1/2 electron, and make it famous. But for some reason Pauli did not write a *three-component* wave, or predict all the polarization classes.

It is possible that Darwin rejected spin-1/2 polarization because it was 'too weird'. The fact this geometrically novel class existed was known by mathematicians before 1900. Under a rotation about an axis by angle $\vec{\theta}$, the spin-1/2 polarization rotates by functions of $\vec{\theta}/2$. We use 'rotates' to mean a literal rotation, as so familiar with vectors, but in a novel geometrical class described by half angles. That is strange, but consistent! The mathematics of the *rotation group* finds and characterizes all possible classes.

Unfortunately the botched old presentations have many basic facts wrong, so *watch out!*. The first mistake fails to recognize polarization at all, and talks about spinning *parcles*. The next gaffe bases almost all the analysis on the spin-1/2 case, probably because it (alone) tends to defy an intuitive picture. Finally, false information is given that 'the presence of $\hbar$ shows that spin is an entirely quantum mechanical effect', which can't be correct when you know the math pre-existed, and predicted all the polarization categories. By 1937 Cartan complained about how physicists abused the concepts, writing in his book *Theory of Spinors* that:

> 'In almost all these (physics) works, spinors are introduced in an entirely formal manner, and without any intuitive geometrical significance: and it is the absence of geometrical significance which has made ...(physics)... so complicated.'

It was the physicists, not the mathematicians who made spin overly abstract.

10.3.2 Polarization observables

As always, the observables of polarization are sums of sandwiches $\psi_j^* ... \psi_k$, where ... stands for summing with operator matrix elements *jk*. There are[11] N^2 combinations of *jk* taken all independent ways. *Spin* is a label for a special set of three matrix operators called $\vec{S}$, which mathy craftsmanship has selected from the N^2 possibilities. Those selected operators project out the actual, physical angular momentum carried by the polarization when $<\psi|\vec{S}|\psi>$ is computed.

Here is an example with a 3-vector wave. Let $\vec{\psi}(t) = (\vec{q}(t) + \imath\vec{p}(t))/\sqrt{2}$, where $\vec{q}(t)$ and $\imath\vec{p}(t)$ are real valued. Let the matrix elements $S_{jk}^{\ell} = -\imath\epsilon_{jk\ell}$, which is the Levi-Civita array that produces ordinary cross products, thus $\sum_{jk} A_j S_{jk}^{\ell} B_k = (\vec{A}^* \times \vec{B})_\ell$. Then

$$<\psi|\vec{S}|\psi> = \vec{q} \times \vec{p},$$

which is the usual formula for angular momentum. It's no accident that symbols $\vec{q}$ and $\vec{p}$ worked out: it's basic craftsmanship.

It is intelligent to make a basis of eigenstates using a nice Hermitian operator. The eigenvalue equation $S_z|m> = m|m>$ has N orthonormal solutions for an N-dimensional case. That basis spans all linear combinations of polarization, so one can use it freely. According to the eigenvalue postulate a measurement of S_z will *always* produce an eigenvalue and eigenstate. (Never say *always* about an experiment; always say *sometimes* that works out to be true!) N spin-labels are not enough, however, to account for the N^2 possibilities in every form of $\psi_j^* ... \psi_k$ that can occur.

[11] While complex ψ_j doubles the counting, $\psi_j^* ... \psi_k$ is Hermitian, so N^2 independent elements is the total.

The upshot is that *polarization is the physical degree of freedom, from which spin-observables are projected,* while *spin is an incomplete attribute that cannot fully describe polarization.* This is not brought out by studying the electron's special case of two polarizations: that's why most material about quantum spin deals with little else—and avoids the word 'polarization'. To master these concepts, you really must study *the rotation group* with all its consequences.

Example: sunlight. You have a beam of sunlight, and measure the energy flux with a thermometer painted black, observing the rise in temperature as a function of time. Does the formula $E = \hbar\omega$ apply? The somewhat educated person can say: 'Yes, since light is a stream of photons. You multiply the number of photons per second by $h\nu$ energy per photon to compute the energy absorbed per second'. That answer is circular to assume and maintain the misrepresented older idea of photons. There is a better answer: 'Why bring up photons?' The experiment is not using a photon counter where the concept would be relevant. The postulate that every measurement of energy (whatever that means) yields an eigenvalue failed: the experiment is an improper measurement, so perhaps should be ignored. Really?

The photon idea was misunderstood and misrepresented to be a universal fundamental entity for every explanation. The picture that light is a stream of photons uses it. It later turned out that a photon is a special *exceptional* state of quantum electrodynamics—an eigenstate—with very limited scope. Lots of electromagnetic states are different. The theorists defined the eigenstates, and by definition every photon has a definite *polarization.* One then asks, exactly what is the polarization direction of sunshine? A student may say: 'Unpolarized light vibrates in all directions at once.' OK, please express that with a mathematical formula. The student realizes that a vector 'in all directions at once' produces a formula for *zero.* The thermometer does not read *zero.*

The unpolarized electromagnetic field is an example of a state not capable of being described with one wave function, and needing a density matrix. In general, *polarization* describes the features of a density matrix. The attempt to dumb it down simply fails in general.

10.3.3 Many observables cannot be expressed with wave functions

Our definition of an observable $<A> = <\psi|A|\psi>$ was more general than the 'observables' dictated by the eigenvalue postulate, which (ok, ok) represents a special case. Beginning with section 2.3.3 we had mentioned that entanglement and the *density matrix* create the general structure of quantum theory. Now we can see why.

Let $|x>$ describe an ideal light wave with electric field polarized along the x axis. By the Born rule, the probability this can be described by $|y>$, representing an electric field polarized along the y axis, is $|<x|y>|^2 = 0$. Suppose the beams labeled 1, 2 have passed through a calcite crystal, with direction $|1>$ strictly correlated with $|x>$ and direction $|2>$ strictly correlated with $|y>$. The normalized state is

$$|\Psi> = \frac{1}{\sqrt{2}}|x>|1> + \frac{1}{\sqrt{2}}|y>|2> .$$

The probabilities (relative intensities) of each beam given the state are

$$P(|x>|1>|\Psi>) = \left|\frac{<x|<1|x>|1>}{\sqrt{2}}\right|^2 = 1/2;$$

$$P(|y>|2>|\Psi>) = \left|\frac{<y|<2|y>|2>}{\sqrt{2}}\right|^2 = 1/2.$$

Adding the probabilities gives the total intensity '1' relative to the incoming intensity. This would be considered obvious. It is not obvious.

The human eye is generally insensitive to polarization. Then ignoring polarization, how do we describe the simultaneous observation of two beams in two directions, as in figure 10.7? A student misunderstanding the 'principle of superposition' would say: 'You add the amplitudes $|1>+|2>$ and square to find the intensity'. That's a common mistake and does not describe what happens. If the two beams are combined, the rule of adding intensities is different from the rule of adding amplitudes, and generally contradicts it. Let's repeat this: ten thousand times a day, students are told to add the amplitudes and square, and never add the intensities. Yet *in many cases you should add intensities* and few seem to notice!

Let A be an operator on the joint space of directions and polarizations. Suppose like the human eye, the A-observables know nothing about polarizations. Then A acts like the unit operator on polarizations, doing nothing to them. Specifically

$$<x|A|x> = <x|x> = 1, \quad <x|A|y> = <x|y> = 0, \quad <y|A|y> = <y|y> = 1;$$

$$<A> = <\Psi|A|\Psi> = \frac{1}{2}(<1|<x| + <2|<y|)A(|x>|1> + |y>|2>)$$

$$\text{or } <A> = \frac{1}{2}<1|A|1> + \frac{1}{2}<2|A|2>.$$

This is just the average of two experiments using direction wave functions $|1>$ or $|2>$ separately, as if polarization never existed. It is *much different* from adding $|1>+|2>$ and computing the sandwich, which would introduce cross-terms of the form $<1|\ldots|2>$.

Suppose a detector can distinguish polarizations, but has no information about beam directions. The corresponding operator B acts like the unit operator on beam directions, doing nothing to them. Then

$$<1|B|1> = <1|1> = 1, \quad <1|B|2> = <1|2> = 0, \quad <2|B|2> = <2|2> = 1;$$

$$<B> = <\Psi|B|\Psi> = \frac{1}{2}(<1|<x| + <2|<2|)B(|x>|1> + |y>|2>)$$

$$\text{or} \quad <B> = \frac{1}{2}<x|B|x> + \frac{1}{2}<y|B|y>.$$

This is just the average of two experiments using polarization wave functions $|x>$ or $|y>$ separately, as if distinct directions never existed. It is *much different* from adding $|x>+|y>$ polarization, which would describe a state of definite polarization direction. Averaging the intensities *as if* they came from equal probabilities of orthogonal components is how one computes a *completely unpolarized* process.

There's nothing special about two states. Suppose you have *a model* where five states of $|luck^{(\alpha)}>$ are strictly correlated with five states of being $|rich^{(\alpha)}>$. Those joint rich-and-lucky states[12] can be written

$$|\Psi> = \sum_{\alpha}^{5} |luck^{(\alpha)}> \Lambda^{(\alpha)} |rich^{(\alpha)}> .$$

If none of your instruments detect inherited wealth, observables will take the form

$$<A> = \Lambda^{(1)} < luck^{(1)}|A|luck^{(1)}> + \Lambda^{(2)} < luck^{(2)}|A|luck^{(2)}> + \cdots \Lambda^{(5)} < luck^{(5)}|A|luck^{(5)}> .$$

The expression is indistinguishable from a weighted sum of observables occurring by pure *luck* with probabilities $\Lambda^{(\alpha)}$. But was every single event necessarily an eigen-lucken-state? We doubt it.

The examples should not be imagined exceptional. Instead they are inevitable. When dealing with a *physical* quantum system, the extent of its *entanglements* will initially be *unknown*. You may believe the system is described by a function of $\vec{x}$. All other attributes and variables not specifically described will effectively be summed out and ignored in the observables you happen to consider. *Reduction* is the process of systematically decreasing the dimensionality of description, in anticipation that some dimensions will not be measured. The *density matrix* is a tool to automate that process and those calculations. If you are good with linear algebra you will recognize that every example was consistent with $<A> = tr(\rho A)/tr(\rho)$ for some matrix ρ, and where tr stands for the *trace*. This definition of the *observable* suffices to define the density matrix ρ. When A are Hermitian, then ρ is Hermitian, and has positive eigenvalues: nothing more is generally required. By a math identity, $tr(\rho A) = <\psi|A|\psi>$ if and only if ρ is so simple it has one eigenvector, $\rho|\psi> = |\psi>$. *Systems that are so simple the density matrix reduces to a single eigenfunction are the exceptions.* Naturally more information is needed to understand the density matrix unless you are *very very* good with linear algebra: that's the next step in learning quantum mechanics in any event.

The opposite happened in early discussions about quantum probability, which were entirely phrased in terms of wave functions. That is because the density matrix (although apparently recognized in some form in 1927) was not understood before John von Neumann's book [3] appeared in 1932. This was about four years after Bohr and Born, Heisenberg, and Jordan (*BHJ*) on one side, and Schrödinger (with help from Einstein) on the other side, had staked out their academic fighting territory. Schrödinger and Einstein absorbed entanglement and emphasized it: Bohr and *BHJ* went backwards and emphasized Planck's constant.

[12] The whole point of being lucky is to be rich enough to do something with it.

The density matrix does not make the Schrödinger wave function any more or less an element of 'reality'. Instead, the density matrix disgraces those who were claiming lofty postulates about what existed in reality. What does it say if your schoolbook is loaded with postulates about the wave function[13] and fails to tell you that no single wave function generally exists?

Due to entanglement the mere *description* of a realistic-quanty-reality is very complicated, whatever that means! The density matrix made the statistical foundations of quantum theory rather clear. *Whether or not* probability is intrinsic, *the Universe is so complicated by entanglement and interactions* that every attempt to describe it uses a density matrix summing over many undescribed and undetected features. There is no such thing as an isolated 'free electron'. Whatever is meant by an electron is immersed and interacting with the myriad variables of quantum field theory, churning and time evolving in the background, with all their initial conditions, and ignored in defining the 'electron'. This is an acceptable explanation for why quantum mechanics makes statistical predictions. *What else would you expect?*

10.3.4 Mott's particle detector and decoherence

When people hear there's no experimental evidence for particles, they can't believe their ears, because their eyes have seen nothing else about 'particle physics' except pictures of little particles exploding in all directions. But think again: when anyone looks closely at those tracks, perhaps using a crystal capable of probing its wave character, they find out the thing is a little wave. It is not mysterious for a little wave to zoom along and be misidentified by eyes, ears and word abuse as a 'particle'.

However, it is quite mysterious for a *spherical wave* predicted by the quantum wave theory to propagate over a macroscopic distance, and resolve itself into little localized packages the experiment indicates. This amounts to one [4] mystery not explained so far. We mentioned it in Section 1.0.4. Meanwhile a million other mysteries were quickly explained by the Schrödinger equation. It was quite unbalanced for 'wave particle duality' to focus on one single particle-like phenomenon, naturally a favorite, and declare that the 'particle aspect' was on an equal footing with the 'wave aspect'. The assertion of complete symmetry of explanations was never true, and that's why it caught on! Yet nothing was ever calculated from the particle aspect. All those things supposedly particle-like were just wave energy and wave momentum things that did not really come from particles. That matters!

In 1929 Charles G Darwin (physicist-grandson of the naturalist) saw the key. A particle track in a cloud chamber is a series of correlated atomic events. It's quite ordinary for a fast spherical wave to pass an atom, and cause a localized, resonant atomic transition to an excited state, triggering a droplet in the chamber. The unexpected thing from a spherical wave is two or more events on a straight line path, rather than distributed over a spherical surface. Yet would the first interaction not

[13] The influence of the Copenhagen presentation led to less and less mention of the density matrix in books written between 1940 and the year 2000. It is mentioned in one phrase by *Kroemer* and not mentioned at all by *Griffiths*. Just as quantum mechanics is having a revival, so is the recognition of the density matrix.

distort the propagating wave, and perhaps lead to favoring more interactions with any subsequent atom on a straight-line path? And since there is an interaction, should the system of a fast wave, plus 1, 2, ...N atomic detectors, not be treated as an entangled whole, or 'multi-verse' of possibilities, impossible to reduce to one single wave? As Darwin wrote [4]

> 'In this "multi-infinite" space a single point then represents the state of the world and a single line its history. It is interesting to observe that the quantum itself plays no part at all in the sub-world, except as a dimensional constant which can be incorporated with the other dimensional constants of the system.'

Darwin's term 'the quantum' refers to Planck's constant. Noticing that it no longer played a role in quantum theory was perceptive, but got him no points from the cult of the pre-quantum theory so desperate to keep it around.

Neville Mott was a young theorist who picked up Darwin's clumsy attempt at calculation, and made it into the first convincing use [5] of entanglement leading to 'decoherence'. The term 'coherence' means one can deal with an isolated wave amplitude using the rules of wave calculations, the Born Rule and so on. 'Decoherence' means that the interactions with a larger system cannot be reduced to something so simple. One either treats the large system as a giant whole, or deals with interactions that erode or destroy the picture of an independently propagating wave. Mott finessed his calculation by choosing the steps needed to make the outcomes plausibly come out. It was ignored, as far as we can see, for nearly 40 years, despite the fact that Mott later received a Nobel Prize (1977) for unrelated work on semiconductors. The literature of quantum mechanics after the year 2000 has far more references to *decoherence* than to *coherence*. It's a hot topic, 75 years in the making!

Why was Mott's contribution ignored? Who benefitted? We can imagine a critic saying that the calculation was not quite based on first principles, but needed to slip in an assertion that certain scattering amplitudes added coherently in a certain way, and incoherently otherwise. Actually, that's fair and necessary. The engineering of a particle detector is hardly a trivial, first –principles affair. The technical experts fiddle endlessly rejecting things that don't leave particle-like tracks, until by a circular process they produce a device that does leave satisfying particle-like tracks.

Actually, quite a few experiments in quantum mechanics have a similar feature. The experiments that don't work out are often set aside as useless until they confirm a theory that gives them value. That was the case with Davvison and Germer, the discovery of the positron, the discovery of Cherenkov radiation, the discovery of parity violation... Isn't it interesting that most discoveries cannot be made until it is the right time for the discovery to be made!

10.4 Bell inequalities, EPR, and all that

In 1964 John Bell published [6] 'On the Einstein Podolsky Rosen paradox', which dramatized what Einstein, Podolsky, and Rosen (*EPR*) had found 30 years before in

Physical Review, 1935. The issue is again the surprises of *entanglement*. Coming a few years after Schrödinger's cat, Einstein had the same intentions to expose and defeat an unscientific, mystical attitude that had taken over quantum mechanics. Already by 1928 he had written to Schrödinger:

> 'The Heisenberg–Bohr tranquilizing philosophy—=or religion?—is so delicately contrived that, for the time being, it provides a gentle pillow for the true believer from which he cannot very easily be aroused. So let him lie there.'

10.4.1 A long and dirty turf war

Jean Bricmont has written [7]:

> 'The history of quantum mechanics, as told in general to students, is like a third rate American movie: there are the good guys and the bad guys, and the good guys won.
>
> The good guys are those associated with the 'Copenhagen' school, Bohr, Heisenberg, Pauli, Jordan, Born, von Neumann among others. The bad guys are their critics, mostly Einstein and Schrödinger and, sometimes de Broglie. The bad guys, so the story goes, were unwilling to accept the radical novelty of quantum mechanics, either its intrinsic indeterminism or the essential role of the observer in the laws of physics that quantum mechanics implies… the views of Albert Einstein, John Bell and others, about nonlocality and the conceptual issues raised by quantum mechanics, have been rather systematically misunderstood by the majority of physicists.'

The assassination of Einstein's character began with disinformation that he stupidly opposed its statistical predictions. Meanwhile, goes the *dis*, Bohr and Heisenberg supposedly revolutionized liberal thinking. Scholars find the opposite: Arthur Fine [8] writes:

> 'It was Bohr who balked at the idea that one might give up the classical concepts and it was then Bohr who worked out the method of complementary descriptions in order to save these very concepts. This is the method that Einstein castigates as a 'tranquilizing philosophy'. Thus the tale of Einstein grown conservative in his later years is here seen to embody a truth dramatically reversed. For it is Bohr who emerges the conservative, unwilling (or unable?) to contemplate the overthrow of the system of classical concepts and defending it by recourse to those very conceptual necessities and *a priori* arguments that Einstein had warned about in his memorial to Mach. Whereas, with regard to the use of classical concepts, Einstein's analytical method kept him ever open-minded, always the gadfly who would not be tranquilized.
>
> In the end Einstein was more radical in his thinking than were the defenders of the orthodox view of quantum theory, for Einstein was convinced that the concepts of classical physics will have to be replaced and not merely segregated in the manner of Bohr's complementarity.'

Isn't that interesting! In their construction Bohr and Heisenberg always retreated to interactions of a classical *parcle* picture, which were never actually computed, and supposedly made quantum systems inherently unobservable. As Yu Shi expresses it [9]:

> 'The Copenhagen interpretation is based on a misinterpretation of the uncertainty and uncertainty relation, and confuses entanglement with local interaction.'

H D Zeh has written [10]:

'The Copenhagen interpretation of quantum theory insists that the measurement outcome has to be described in fundamental classical terms rather than as a quantum state .... The Copenhagen interpretation is often hailed as the greatest revolution in physics, since it rules out the general applicability of the concept of objective (classical) physical reality. I am instead inclined to regard it as a kind of 'quantum voodoo': irrationalism in place of dynamics.'

We inserted the word *classical.*

John Wheeler called the mystical voodoo stuff 'quantum mumbo-jumbo'. Among other things, Wheeler was indirectly referring to J R Oppenheimer, his colleague at Princeton, and known as 'Oppy'. E T Jaynes later became one of the 20th Century's most insightful writers on probability in physics. He wrote about his interactions with Oppy as a student[14]:

> 'When in the Summer of 1947 Oppy moved to Princeton to take over the Institute for Advanced Study, I was one of four students that he took along .... My thesis was to be on Quantum Electrodynamics .... But, as this writer learned from attending a year of Oppy's lectures (1946–47) at Berkeley, and eagerly studying his printed and spoken words for several years thereafter, Oppy would never countenance any retreat from the Copenhagen position, of the kind advocated by Schrödinger and Einstein. He derived some great emotional satisfaction from just those elements of mysticism that Schrödinger and Einstein had deplored, and always wanted to make the world still more mystical, and less rational .... Some have seen this as a fine humanist trait. I saw it increasingly as an anomaly—a basically anti-scientific attitude in a person posing as a scientist—that explains so much of the contradictions in his character .... I wanted to reformulate electrodynamics from the ground up without using field quantization. The physical picture would be very different; but since the successful Feynman rules used so little of that physical picture anyway, I did not think that the physical predictions would be appreciably different; at least, if the idea was wrong, I wanted to understand in detail why it was wrong .... If this meant standing in contradiction with the Copenhagen interpretation, so be it; I would be delighted to see it gone anyway, for the same reason that Einstein and Schrödinger would. But I sensed that Oppy would never tolerate a grain of this; he would crush me like an eggshell if I

[14] Jaynes switched in 1948 to Eugene Wigner as his thesis advisor.

> dared to express a word of such subversive ideas. I could do a thesis with Oppy only if it was his thesis, not mine.'

Why was Bohr so successful? It is reported that his lectures were mesmerizing philosophy. They were based on few equations, which was attractive. It is reported Bohr got everything from the de Broglie relations, and classical particles. And Bohr commanded tremendous financial resources from endowments and the support of the Danish government.

We must not forget there was a world war from 1914 to 1918. There were catastrophic consequences for Germany after 1918 from the Versailles Treaty, the ruin of the world economy by the stock market crash of 1929, and a depression in Europe so deep it made the out-of-work Americans seem rich. (The Americans usually had food.) Theoretical physics has always demanded courage: everyone attempting it after 1914 was a special kind of reckless fool. Yet everyone needed food. Visiting Bohr for a spell might yield the equal or better of a year's salary, when people were living month to month. If you were his assistant your future was secure. Victor Weisskopf was an assistant to Bohr for a few years, and recalled the following[15]:

> 'It is very difficult to get into Copenhagen; I have seen cruel things happen if you come and cannot get through the 'Guard'. Bohr was surrounded by five or six, maybe even more, of his disciples, who were a very arrogant crowd. If you were not accepted by them you would have a very difficult time with him. That was always so, and I can give you a few examples. Rabi is one; a number of Americans had a very bad time here .... They would see Bohr very little because we watched it. I know because I was one of those disciples; we were not nice. Well we did it out of tremendous enthusiasm, to keep the level high.'

During World War II Werner Heisenberg was the physicist leading the German Nazi government program to develop an atomic bomb. The United States had its own program, initiated with worry that Germany might succeed first, and sent spies to monitor progress. One of the spies was former baseball player Moe Berg. His baseball career (1926–34) had not been spectacular, except that he was a noted intellectual. Sports writers quipped[16] that: 'He can speak 12 languages but can't hit in any of them.' In 1944[9] Berg was in Germany with instructions to gather information on the bomb program. He knew enough physics to befriend top theorists and be invited to attend lectures. In 1944 he attended a talk by Heisenberg with a pistol in his pocket[17] 'with orders to shoot Heisenberg if his lecture indicated that Germany was close to completing an atomic bomb. Heisenberg did not give such an indication, so Berg decided not to shoot him, a decision Berg later described as his own 'uncertainty principle'. Berg's keenly perceptive observations of Heisenberg and expertly extracted information may

[15] https://www.aip.org/history-programs/niels-bohr-library/oral-histories/4944

[16] http://www.baseballlibrary.com/ballplayers/player.php?name=Moe_Berg_1902

[17] The phrase in quotes comes from *Wikipedia* and is so well done we left it alone. But we're not perfectly sure what Moe Berg actually said.

Figure 10.8. Former major league baseball player Moe Berg played a decisive role in physics; see the text.

have influenced the decisions of President Roosevelt and General Groves, the US military man heading the Manhattan Project, *not to assassinate* Heisenberg along with the entire community of German quantum physicists.

After the war Moe Berg refused to accept a national medal for his heroism. At some point someone criticized him for 'wasting his intellectual talent' on baseball. Berg (who had a degree in law) replied[16]: 'I'd rather be a ballplayer than a justice on the U.S. Supreme Court'.

10.4.2 The *EPR* allegory

Plato had an allegory of the cave. It was about people kept in ignorance, and given partial information from the movement of shadows. *EPR*'s allegory of quantum mechanics was meant to challenge those presenting shadowy pictures involving the uncertainty relation, *parcles*, the 'collapse postulate', and *locality*. *EPR*'s strategy [12] was war-like use of the Copenhagenish-($\mathcal{C}$) methods and rules against itself.

Einstein was never insensitive to language: he knew perfectly well that repeatedly mismatching quantum *parcle* words to calculations was the war tactic of his enemies. Then *EPR* describe an example they know very well does not exist, and where time evolution, the only prediction of physics, deliberately plays no role. Consider[18] a certain wave function $\psi(x_1, x_2) = \delta(x_1 - x_2 - x_0)$. It cannot actually

[18] One should read the original paper, which makes an argument more general than a single example.

exist, but it will be said to 'describe two *parcles*'. Let a $\mathcal{C}$-measurement of x_1 produce eigenvalue x_1. The measurement predicts the number $x_1 - x_0$ is then an eigenvalue of the operator $Q = x_2$. Thus, 'collapsing' the wave function at one location collapses it to Q at another, distant location, without any interaction at the other location. That seems physically wrong, but continue. Suppose the momentum of particle 1 is measured. By Fourier transform,

$$\psi(x_1, x_2) = \int \frac{dp}{2\pi} e^{\imath p(x_1 - x_2 - x_0)}. \tag{10.5}$$

The $\mathcal{C}$-measurement of momentum yields an eigenvalue p_*, collapsing system 2 to wave function $e^{\imath p_*(-x_2 - x_0)}$, which is an eigenstate of $P = -\imath\partial/\partial x_2$ with eigenvalue $-p_*$. Once again, the collapse of one system collapses the distant system without interaction. '*Thus, it is possible to assign two different wave functions (in our example* ψ_k *and* ϕ_r*) to the same reality*', wrote *EPR*. (Their ψ_k *and* ϕ_r referred to a general discussion coming before their main example.) Now, argue *EPR* in a subtly flawed way, the sharp position Q and momentum P values of system 2 violate the uncertainty relation, (or better) violate the rule that operators which do not commute cannot have joint eigenvectors. They write:

> 'We are thus forced to conclude that the quantum-mechanical description of physical reality given by wave functions is not complete .... On this point of view, since either one or the other, but not both simultaneously, of the quantities P and Q can be predicted, they are not simultaneously real. This makes the reality of P and Q depend upon the process of measurement carried out on the first system, which does not disturb the second system in any way. No reasonable definition of reality could be expected to permit this.
>
> While we have thus shown that the wave function does not provide a complete description of the physical reality, we left open the question of whether or not such a description exists. We believe, however, that such a theory is possible.'

The reader can consult the original paper for what is 'reality'. The usage connected to equations is that if 'the physical quantity A has with certainty the value a whenever the particle is in the (eigen) state given… (then) there is an element of physical reality corresponding to the physical quantity A'. The purpose of this gambit is to have the claimed contradiction of joint eigenvectors violate reality. For the meaning of a 'complete description', *EPR* write:

> 'Whatever the meaning assigned to the term complete, the following requirement for a complete theory seems to be a necessary one: *every element of the physical reality must have a counterpart in the physical theory.* We shall call this the condition of completeness.'

Considered soberly, the business of making postulates about 'reality' on the basis of an oversimple description was a bit preposterous. Einstein later regretted the presentation which (in our opinion) too much emulated Bohr's pompous style of obscurantism. So what did *EPR* accomplish? First, their wave function $\psi(x_1, x_2) = \psi(x_1 - x_2)$ does not describe two independent particles, and they seem to know it: it is *deliberately and by construction incomplete,* defining a system lacking much information except for a correlation by *entanglement.* (For example, the 'center of mass' of the system $x_1 + x_2$ is 'everywhere',). The Schmidt decomposition is explicit in the diagonal sum of equation (10.5). Since the wave function does not predict much about either system, kinematic assumptions made for independent systems don't apply. The adroit subversion of the collapse postulate demonstrated its lack of predictive power, unless circularly true. What *EPR* accomplished was proof that Einstein not only appreciated quantum mechanics, but had studied it more deeply than his opponents.

Bohr announced to the world there was 'nothing new' in *EPR* and that he had refuted their argument. Actually the rambling seven-page paper reiterates Bohr and Einstein's old and stale debates about the uncertainty relation. Where Bohr appears to respond to the *EPR* content, he wrote:

> 'We see that the argumentation of the mentioned authors does not justify their conclusion that quantum-mechanical description is essentially incomplete. On the contrary this description, as appears from the preceding discussion, may be characterized as a rational utilization of all possibilities of unambiguous interpretation of measurements, compatible with the finite and uncontrollable interaction between the objects and the measuring instruments in the field of quantum theory.'

John Bell wrote[19]:

> 'Indeed I have very little idea what this means .... I do not understand the final reference to 'uncontrollable interactions between measuring instruments and objects', it seems just to ignore the essential point of EPR that in the absence of action at a distance, only the first system could be supposed disturbed by the first measurement and yet definite predictions become possible for the second system.'

10.4.3 Bell physics

John Bell recognized the *EPR* paper had exposed features of entanglement that the $\mathcal{C}$-presentation unwisely downplayed. Bell made his own example of non-local correlations that included a viable experimental prediction. *EPR* has been repeatedly reported to favor 'classical hidden variable theories': you won't find that in *EPR*, while it comes from associating them with Bell.

The standard notation begins with a certain two-component wave function $|\psi^0>$, which can be written as

[19] The quote and context come from Bricmont [7]

$$|\psi^0\rangle = \frac{1}{\sqrt{2}}(|+a\rangle\,|-b\rangle - |-a\rangle\,|+b\rangle),$$
$$\psi^0\rangle = \frac{1}{\sqrt{2}}(|+\rangle|-\rangle - |-\rangle|+\rangle). \tag{10.6}$$

Here $\pm$ stand for the $\pm$ eigenvectors of a matrix σ_z on either space. Experts call the wave function 'the antisymmetric singlet'. The second form omitting system a, b labels is the most common. Already an invisible notational barrier is being set up, using the $|>|>$ notation lacking the detail to be self-explanatory.

We prefer index notation. Let system a have index j, and system B have index K: upper and lower case indices will help distinguish them. Any two-index space has a special antisymmetric matrix

$$\varepsilon_{jK} = \begin{pmatrix} 0 & 1 \\ -1 & 0 \end{pmatrix}.$$

This matrix does not change under any unitary linear transformation rotating the spaces labeled j and K. That is quite special: in fact, it is the only combination of products which is *invariant*. The unit matrix δ_{jK} and the Levi-Civita array $\varepsilon_{jK\ell}$ are prototypes of the same invariant phenomenon in three dimensions. From the invariance $\sum_{jK} a_j^* \varepsilon_{jK} b_K$ does not change under rotations. You can consider it to be the z component of a cross product in the plane, which is nearly the same thing, but not as sharp.

Then IF you are thinking about two electrons, with polarization j K indices, someone plays a dirty trick to introduce Bell's entangled combination:

$$\Psi_{jK}(t) = \varphi(t)\varepsilon_{jK}. \tag{10.7}$$

This is just the same as equation (10.6). The trick is that the array Ψ_{jK} appears to describe four complex numbers (eight facts!) in $j = 1, 2$ and $K = 1, 2$ combinations. However, ε_{jK} is a privileged, invariant combination that is a unit of concept all on its own. There is only one undetermined complex number $\varphi(t)$ in $\Psi(t) \sim \varphi(t)$. One complex number severely lacks the information to describe two spin-1/2 states. Since one overall constant drops out of a quantum state, the wave function shown represents almost, but not quite, zero information. This is very well hidden in the expression or equations (10.6) or (10.7) so few notice. Please notice that we begin with an initial condition having almost, but not quite, zero information. Note this! The state in question is the same state no matter how the polarizations are rotated. It does *not* describe two happy independent little jelly bean electrons. Yet in the set-up, there are always statements the wave function is the initial state of 'two quantum particles', from (say) the decay of a 'virtual photon' to an e^+e^- pair[20]. The imagination by word abuse of the experiment does not correspond to what is underway.

Next in the story-telling, 'the quantum particles explode in opposite directions', moving like Newtonian bullets toward fast detectors placed on the left and right.

[20] A virtual photon has one unit of angular momentum, not zero, but it's a thought experiment!

This is *not* described by the wave function shown: in fact, most discussions even omit the label t for time evolution. Yet the exploding quantum particles are very real in the imagination of the reader being set up for a surprise. Then discussions continue to detecting $\pm$ spin states on the left, and $\pm$ spin states on the right. Here is the eigenstate postulate. If those detectors are Stern–Gerlach instruments, they can still be oriented at separate angles θ_L, θ_R on right and left, leading to many interesting possibilities. Many of the outcomes are *strictly correlated* between both sides. Consider aligned detectors $\theta_L = \theta_R$ and assume no intermediate effects of time evolution between initial state and detection. Then there are no cases observed with -left AND -right. The observable for that case is

$$<\Psi|-L> \mid -R> <-L|<-R|\Psi|>.$$

Inserting Ψ from equation (10.6) will instantly confuse you: the notation is not at all self-explanatory, which the experts enjoy. With indices any case where state j and K match is proportional to δ_{jK}. The inner product $\sum_{jK} \delta_{jK}\varepsilon_{jK} = 0$. The observables for (−left AND −right), OR (+left AND +right) are *zero.*.

That leaves (−left AND +right) and vice versa to observe. In the next step, various misinterpretations claim it is *really weird and incredible* that a particle on the left would show such a 'response' to distant particles on the right. To make this seem weird there's a preamble hinting that getting + on the right, say, must need some *non-local* interaction to force—to happen on the left. The hint is enlarged to discussions about measurements made so quickly it is impossible for communication within the limitations of light-speed between the two sides to be possible. The words 'quantum non-locality' refer to *EPR*'s suggestion there was something wrong with distantly separated systems being correlated.

We'd like you to understand the argument, not be victimized by suggestive language. Pause to consider. The premises of the argument were that the initial state is *not* a simple product of two independent systems. Indeed the math says there are not two independent systems: the wave function never described that. The premises define a state with a *correlation* encoded in entanglement. If you begin with correlations, and make a rule they are preserved over long distances, your own assumptions predict long-distance correlations. To make a fake mystery of this, you're supposed to forget the correlations. In some presentations the 'quantum particle on the left' is presented as an independent entity, just born out of the quantum egg, and minding its own business. To emphasize that (wrong) idea, a side calculation will find that $\pm$ cases happen on the left side exactly as if the state is *unpolarized.* We are supposed to forget that *correlations* between subsystems cannot be found by evaluating subsystems one by one on their own.

Now that we have your attention on *correlations*, recall that by rotating the left and right detectors to angles θ_L, θ_R many more outcomes can be explored. With the restrictive assumptions of the eigenvalue postulate, the eigenvalues measured on left and right will make a table of outcomes where all the information is in the correlations of the left and right sides, which for discussion might be a thousand miles apart. The idea that a quantum particle a thousand miles from another has a life and existence on its own is very appealing. Both the *EPR* and Bell discussions are set up to cause a conflict with that presupposition of 'locality', which is then contradicted by the quantum mechanical premises.

Here are a few analogies to guide the thinking.

The exploding Planck mill One day a classical lumber mill exploded, and sent boards emerging at relativistic speeds which hit unlucky people in neighboring counties. Before the explosion the boards had been neatly stacked in horizontal piles. A supervisor on the left side of the mill was hit in the face by a horizontally flying relativistic plank. That was enough to predict the supervisor on the right side of the mill also got hit with a plank. The two planks were *strictly correlated* in their geometrical orientation. Moreover, the prediction of the double-Planck injury could be made faster than light-speed communication permitted.

A mistaken gem of relativity pop-culture imagines communication at the speed of light to be an issue. But relativity makes no rules restricting what you happen to know in advance. Another wrong gem of quantum pop-culture forgot that correlations built into a system initially will either appear later in time-evolved form, or stay the same. There's nothing remarkable about correlated wood products. *Also*, the instant your model did not find a vertical board in any event, you might predict *with certainty* the wave function for anyone finding a vertical board *collapsed to the outcome of the experiment*. That is the content of the subsidiary 'collapse postulate' that comes with the eigenvalue postulate. It sometimes says that your *description* of a system can collapse instantly. Many instantaneous collapse examples have been set up to dramatize the weirdness of quantum mechanics, while they involve nothing but the collapse of a *description* of a system that was never intended to be complete. The exploding Planck mill explains that parts of quantum measurement paradoxes sometimes have nothing to do with quantum mechanics. But quantum probability is not trivial, and happens to have more in store.

Where is Sally? Since the framework of quantum probability is a mathematical construction, we can't stop any number of applications made to amuse ourselves. In this allegory Sally is a two-dimensional (but complex) person, whose state $(1, 0)$ lives in Lawrence, Kansas, which is a US State, and state $(0, 1)$ lives in Paris, France, which is a nation state. Stranger things have been observed. Today Sam is lonely and uncertain where Sally is. In his mind Sam proposes the Sally state $|Sally> = (1/\sqrt{2}, 1/\sqrt{2})$. It would be an insult to Sally to reduce her to two complex numbers. Also, none of this is to be taken as the actual opinion of Sam, who holds Sally very dearly. It is nothing but *bookkeeping* for certain *correlations*.

What Sam has in mind is a Born rule probability 1/2 that Sally is in either state. Misunderstanding the bookkeeping has led to some fake paradoxes. Light speed communication across the surface of the Earth between Lawrence and Paris has a time delay of about 0.05 s, which is long enough for a human to notice. The door opens, Sam sees Sally, and her wave function collapses from $(1/\sqrt{2}, 1/\sqrt{2}) \rightarrow (1, 0)$ in less time than physical laws should permit. *Did Sally's quantum wave move faster than light?* If Sam thinks so, and finds it mysterious and profound, you should not.

Actually the bookkeeping was inappropriate. Orthogonal vectors should be used for mutually exclusive states. It is true that (1, 0) and (0, 1) are orthogonal and potentially appropriate for objects separated by 8000 miles. Given that the Born probability of (1, 0) is 1/2 Sam made a common error to assume $\psi_{mix} = (1/\sqrt{2}, 1/\sqrt{2})$ to represent his 'state'. Meanwhile $\psi_{antimix} = (1/\sqrt{2}, -1/\sqrt{2})$ also has 50:50 Born rule probabilities to observe (1, 0). They seem both equally good. Yet ψ_{mix} is orthogonal to $\psi_{antimix}$, which has Sam predicting *zero* probability for conditions that are 50:50 Born rule probabilities. Is that a paradox?

It is not a paradox, but a misunderstanding, or inappropriate use of notation. Suppose you know nothing about the elements of an N-dimensional space. You will be wrong to guess $\psi = (1/\sqrt{N}, 1/\sqrt{N}, \ldots 1/\sqrt{N})$. It is a very definite particular sharp state, and predicts $N = 1$ mutually orthogonal vectors will have zero probability, which amounts to knowing quite a bit of information (while you had none). If you have no information, you should not be using a wave function at all.

We cannot develop much of density matrix theory here. Yet we can at least settle the description of a 'completely random system'. Such a system has no preferred directions, and its density matrix is proportional to the unit matrix, $\rho = \frac{1}{N}1_{N\times N}$. There is no information in this and nothing useful to measure. It is as far from defining any wave function as mathematically possible. If you pursue this[21], you will discover that every independent observable operator A has $<A> = tr(A\rho) = 0$.

The upshot is that quantum probability is not the most obvious bookkeeping scheme. It will take more study than this monograph to master it. Speciously assigning wave functions may under-describe what's known, over-describe what's not wanted, or introduce correlations that were not intended. Schrödinger made a parody ridiculing the description of a cat with $(1/\sqrt{2}, 1/\sqrt{2})$. When he introduced the cat he wrote [13]: 'One can even set up quite ridiculous cases', with the intention of highlighting the absurdity of Bohr's views. Schrödinger's enemies then misrepresented it as adopting their views!

10.4.4 Quantum probability is not defined by distributions

Early and often, beginning with section 1.1.2, we warned you that quantum probability will be wrongly defined when presented with $\psi^*\psi$ and statements about distributions. The 1964 paper by John Bell made physics history by dramatizing observables not consistent with distributions. Either Bell already understood that people had been misled to assume distributions, and entertained himself with contradicting it, or he discovered for himself what he already knew. To this day, people misreport and misrepresent the outcome, which is that *if you assume quantum probability and the Born rule come from distributions, it fails in general.* Like many or all 'paradoxes' of quantum theory, it is a completely unremarkable fact that only seems remarkable when you have wrong information.

[21] The trick writes $A = A - tr(A)1_{N\times N}/N + tr(A)1_{N\times N}/N$. The second term is not independent of $1_{N\times N}$, leaving the first term that has no trace.

Letters between Schrödinger and Einstein show they both anticipated that the Copenhagen presentation contradicted itself. Schrödinger attempted to show it by creating observables inconsistent with a distribution, but did not succeed. *EPR* argued that correlations of subsystems at arbitrary distances were inconsistent with an independent 'reality' of the subsystems, which (as *parcles*) the Copenhagen presentation had assumed. Physics waited for Bell to set up a competing model using distributions, and show it would not work.

Bell's approach went as follows. Suppose systems A and B are described by an entangled wave function: the *EPR* example is often used. Suppose a and b are eigenvalues of operators from measuring A and B, as the most restrictive measurements assume. Let $P(a, b|c) > 0$ be a classical distribution of a and b depending on some other random variables c. Add a little detail that when some particular c occurs, it necessarily and always produces some definite outcomes, such as $a \to +$ and $b \to -$. (Many have noticed that Bell sets up a 'straw man' distribution model which is not totally compelling, but continue.) Then find an operator W such that the quantum correlations of $<\Psi|W|\Psi>$ cannot be reproduced by any such $P(a, b|c)$:

$$<W> = <\psi|W|\psi> \overset{!}{\neq} \int dadbdc\ W(abc)P(a, b|c)P(c).$$

The substantial difference between the left and right side, and tremendous freedom to choose $|\psi>$ and W, give every reason to believe that a theorem 'can't always be done' (symbol $\overset{!}{\neq}$) should be possible.

That kind of analysis will be tricky, because none of the cases where an equality is true will help the objective. It's often difficult to express what the infinite powers of math *cannot* do, so that Bell was forced to proceed by algebraically deducing some inequalities obeyed by the right hand side. Then any quantum mechanical $<\psi|W|\psi>$ violating the inequality falsifies the existence of a distribution-based description.

Many sources repeat Bell's calculations, but we find it is the kind of algebraic exercise that happens to teach you very little. You are understanding Bell's accomplishment and message when you understand that *dramatizing* a feature of quantum probability, which was always present in the Born rule, did not discover anything new. An experimental verification of the analysis also did not discover anything new. That showed that entangled states exist, but we already knew that entangled states exist. It is interesting for states to preserve their entanglement over macroscopic distances: and also somewhat of a technical stunt to make it happen.

What was actually new, and got many excited, was the language that it 'rules out the general applicability of the concept of objective (classical) physical reality'. Those words *objective* and *classical* make a difference. Bell knew he had material to excite popular interest, and deliberately oversold the results as showing that 'no hidden variable theory can reproduce quantum mechanics'. Here 'hidden variable' was improperly extended to more than Bell had shown. It was used to mean *any* theory where causal, deterministic physical variables, not themselves statistical quantities, were guided by background variables like c to produce apparently random outcomes. Bell's simple model was not general enough to show that: yet a dumbed-down and sweeping 'no hidden variable' statement became popular anyway. (Circular and true, the kind of hidden variable theories that are not good enough are the particular sort of classical distribution models of the type made by Bell. Quantum mechanics itself is loaded with hidden *quantum* variables that tend to go unnoticed!)

Bell's work was followed by many variations from many authors. They all depended on the breakdown of classical distribution concepts, while usually talking about something else. Many emulated Bell's method by producing inequalities. Inequalities were so popular that *not* using an inequality to express a math fact was news. The abstract of Lucien Hardy's 1993 paper [14] entitled *Nonlocality for two particles without inequalities for almost all entangled states* is:

> 'It is shown that it is possible to demonstrate nonlocality for two particles without using inequalities for all entangled states except maximally entangled states such as the singlet state. The eigenvectors corresponding to the measurements that must be performed to do this are exhibited and found to have a particularly simple relationship to the entangled state.'

'Exhibiting nonlocality' means that one can mathematically define long-range quantum correlations that a model based on distributions of autonomous, local entities cannot reproduce. We doubt that experimental stunts can verify every possible long-range quantum correlation that can be mathematically defined, but there's an active industry in attempting some of it.

10.4.5 Something weird

As Gell-Mann indicated (section 3.2.3) Bohr's last accomplishment with quantum mechanics was writing with exponentially increasing obscurancy so that physicists lost interest in quantum mechanics for 50 years. A 2010 conference[22] was sufficiently brave to advertise: 'Why should young people be interested in these ideas, when showing interest in quantum foundations still might harm their careers?' Yet the development of technology capable of experimentally testing *EPR* and Bell-type correlations has reanimated interest in foundation questions.

All around the world nowadays, experimentalists are entangling systems and astonishing witnesses with the outcomes of *ordinary* quantum mechanics. The lesson every time is that two or more entangled *anythings* are not so many *things* but ONE UNIFIED CORRELATED SOMETHING, so that words like 'photons' fail to work right. Figure 10.9 shows that the human visual system can see rather deeply into entangled functions, and extract information. The un-entangled image in the upper left corner has almost no information.

Figure 10.10 adapted from an experimental paper [15] shows an optical device exploiting pairs of entangled photons. The entanglement is such that those *parcles* (or that UNIFIED THING) is (or are) emitted from a laser-driven crystal in back-to-back pairs, with equal and opposite momenta, just as *EPR* describe. On the left side of the figure a perforated aperture or mask with a pattern is placed in front of a lens and a

[22] https://vallico.net/tti/master.html?https://vallico.net/tti/deBB_10/announcement.html.

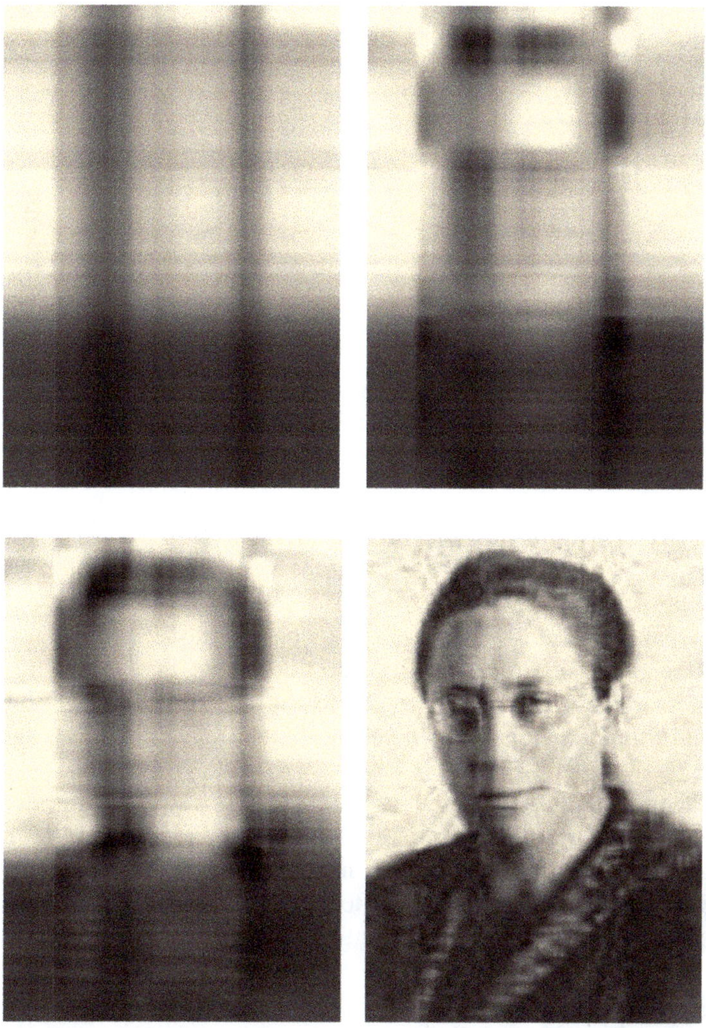

Figure 10.9. A demonstration of the Schmidt or (singular value) decomposition of an entangled function $f(x, y) = \sum_{\alpha}^{\alpha-max} a_\alpha(x)\Lambda_\alpha b_\alpha(y)$. The panels use $\alpha - max$ =1, 2, 5, 20 out of a possible 324. The original image is a 324 × 450 pixel JPG of the brilliant mathematician Emmy Noether (1882–1935).

detector called D_1. It is illuminated by the beam moving to the left. The other beam goes in the opposite direction by a uninterrupted flight in free space. The right side has a movable fiber optic detector D_2, which operates in timing coincidence with D_1. Data from rastering detector D_2 over the transverse plane *on the right hand side* finds a sharp magnified image of the aperture. To be clear, no light goes from the left side, where the aperture is, over to the right side. Yet on the right side, detector D_2 detects the image using entangled THINGS.

One's first impression is that this is impossible magic. The 'ghost image' on the right side somehow knows what the left side saw. This first impression assumes there are two independent, non-interacting photons, which the experimental evidence

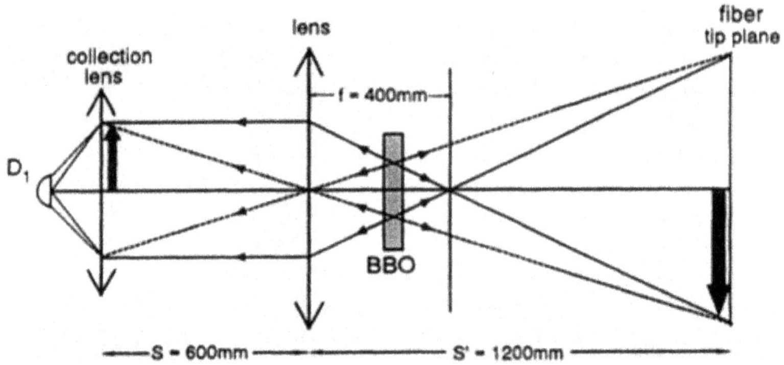

Figure 10.10. Schematic layout of an apparatus producing 'ghost images' by entanglement of photons. A laser-driven crystal emitting back-to-back correlated photons is shown as BBO. A perforated aperture with a pattern is on the left side in front of detector D_1. The photons producing a focused ghost image at detector D_2 is unimpeded, and never goes through a lens.

contradicts. (Removing the coincidence requirement DOES yield random photons on either side, with no ghost image formation.) Yet the next question is whether the data might be explained by classical physics. When the photon on the left goes through any particular point of the mask, a classical model has an exactly oriented photon on the other side to match its direction, event by event. The coincidence detection keeps the events ordered and different directions from getting mixed up. However that forgets there's a *lens* focusing the photons on the left, and *no lens* on the right!

We'll let the reader decide whether a lens on the right *should* be needed. In a sense ghost imaging is not more mysterious than a locksmith's key-cutting machine that reads one template and cuts another in coincidence. Entanglement is *not* a necessary feature [16], and classical signals can do almost the same thing: except for the subtleties of quantum probability of entanglement, which are so interesting.

10.5 Chapter summary

The chapter began with a statement about the description, *sums of products are generic*. If physics had gone differently, a description based on classical probability and its gross over-description of waves might have been the default. After assuming distributions of unimaginable complexity, physicists might be mystified to find patterns of experimental outcomes of rather simple type, like +, +, −, −, +, −, +, −, −, +, ... whose correlations could not be described by distributions. This would be quite surprising, because mathematicians (Kolmogorov) and physicists always assumed classical probability could describe anything that might ever happen, or be *conceived* to happen. If not already known, experiments would have forced scientists to adopt a larger, and more flexible framework for probability, with the misnomer, *quantum probability*. And the subject would not be about quantization.

As history went, there was an early decision to adopt the Born rule to *define the kind of probability used in the description*. The idea of using classical probability for waves was never seriously considered. If you are going to use a wave function (or density matrix) for your description, you will always come in a circle to quantum probability. The lesson of quantum mechanics is that forcing prior opinions onto your description of nature seldom works out. Since wave functions and density matrices parameterize what can be observed, and vice versa—all of it based strictly on what is observable (section 8.2.3)—the framework is flexible enough. We don't know what nature is, and it is not clear whether quantum theory fully describes it. However, *it's not the worst thing. It has not failed yet.*

10.6 Suggested reading

There are many books about quantum mechanics. Here in no particular order are our observations about some of them. Book titles are generally suppressed when author names suffice.

- *Modern Physics* denotes books we must warn you to stop consulting. Contrary to the 'ethical code' of physics teaching, books on modern physics deliberately plant misconceptions and mistakes for reasons of expedience. Books such as *Krane* favor operational authoritarianism: they tell you the formula, and you execute the drill. Books such as *Serway, Moses and Moyer* shoot for sneaky compromises or 'white lies'. If you find a cookie-cutter formula in a modern physics book, its limitations and context will generally be botched, making a recipe for lifetime incompetence. No book entitled *Modern Physics* should be opened after a person opens books entitled *Quantum Mechanics.*
- *Bes* is an undergrad book that is handsomely concise. It is rigidly aligned with the Copenhagen school based on authority statements and eigenvalue equations as first principles. Time evolution is postponed until the book is nearly over. An interesting chapter deals with quantum computing.
- *Griffiths* is an undergrad book on wave mechanics. Griffiths has the highest presentation skills and friendly manners but chooses never to innovate anything. A calculated level of symbol-clutter and operational clumsiness keeps students challenged on the wrong tasks. Some teachers do that deliberately.
- *Ohanian* is an undergrad book-writing professional who is competent enough. An innovative approach of ladder operators unifies the book and solves the hydrogen atom. In order to make a buck some of the material is dumbed-down.
- *Gasiorowicz* is a good book for the transition between undergrad and graduate level. It starts from nothing, but goes the whole way and in many cases is cleaner than Griffiths.
- *Winter* is a delightful, hard to find text by a professor from The College of William and Mary who cared deeply about teaching well.
- *Blümel* is a 2010 book that concisely summarizes the Copenhagen tradition, wave–particle duality, etc, and then spends half the book on modern applications of quantum information that contradict the Copenhagen

tradition. The book is much more thoughtful and complete on measurement theory than others we reviewed.

- *Kroemer* is a marvelous book from a great master of condensed matter physics. There's a good deal of fine original material found no place else. One does notice not much deep thinking about foundations or the order of presentation.
- *Robinett* is a solid serious undergrad book notable for showing the work done in calculations in complete and unpretentious straightforward detail.
- *Chester* is a superstitious little book mixing up basic facts of linear algebra with profound physical principles and sublime mysteries.
- *Goswami* is an undergrad-level book by a dedicated nuclear theorist who tries to balance philosophical aspects. It is one of the few that reviews the struggles, sense, and nonsense of different interpretations of measurement theory while still discussing physics.
- *Bohm* is a relic of more than 50 years ago when great men struggled to make sense of quantum theory. Bohm was philosophical yet critical of the mumbo-jumbo and suspected physics ought to explain physics. The overall level of this critical treatment is a bit math-challenged and below the level of undergrad books by now.
- *Fermi, Notes on Quantum Mechanics* is priceless. These handwritten notes Fermi made preparing for lectures show the amazing simplicity, directness, and efficiency of one of the greatest physicists in history. The only flaw is there's not more of it.
- *Heisenberg* was around at the right time and right place with the masterful skills to produce a masterful book, but it isn't. Nowadays the emphasis on the uncertainty relation as a foundation of physical law appears (to us) deceitful. Certain passages do illustrate the man's adept math skills.
- *Sakurai* is a standard graduate text. Sakurai was a brilliant theorist whose incomplete manuscript was patched together by colleagues after he died unexpectedly. The result notably lacks coherence and unity, which many notice. It has a good treatment of transformation theory, the rotation group, and the other material Sakurai actually finished.
- *Shankar* is an undergrad quantum mechanics book attempting to be a standard graduate text. The methods are usually cumbersome, and concepts generally don't really reach the graduate level. Angular momentum and the rotation group are inexcusably incomplete. The first chapter on linear algebra is excellent and worth the price of the book.
- *Dirac* wrote the classic text early. Dirac's style is marvelous. He links together elegant math, clear physical thinking, arbitrary assumptions, logical gaps, and non-sequiters in a uniform consistent tone. The book is a gold mine of insight about how the tentative and risky early era of quantum mechanics got locked in.
- *Ballentine* is one of very few books by a master who has thought long and deeply to provide a fresh point of view advocating the *ensemble interpretation*. It is so much more coherent than the *C-presentation* that we believe

Copenhagen has already been abandoned (but not the city). The book is complete and appropriate for a graduate-level course.

- *Shiff* was the graduate text of a previous era. The treatment of the student is rather cruel. In derivations less accessible methods are chosen over easy ones, and every sentence is written to make it hard. *Shiff* is like *Jackson* for quantum. Still there is a lot of valuable material.
- *Landau and Lifshitz* is a series across physics representing the ultimate in cruelty to students. Yet Landau was a great physicist and just touching his books is a privilege. The quantum mechanics books are now old-fashioned, and in retrospect naive, given how much Landau knew and grew. The books are useful for obscure insights and many worked problems which tend to be hard. Landau will show off his tricks, but never give them away!
- *Messiah* replaced Shiff with more kindly organization and longer more complete discussions. It has a somewhat mystical character emphasized by profound quotations in French. The treatment of angular moment is good and coverage is vast.
- *Mertzbacher* used to be a text competing with *Shiff* by giving better explanations and more physics. It was greatly expanded through several editions and made more rigid.
- *Commins*' self-described *experimentalist's approach* seems to lack the critical thinking and disbelief of specious theoretical claims we thought experimental physics was all about. Yet the author was a distinguished experimental physicist with many awards. The book seeks to be a monumental summary of all physics from zeroth level to electroweak quantum field theory.
- *Liboff* is a stuffy unprogressive text with a level too high for an introduction and too low for graduate courses. 'Uncertainty as the foundation of natural law' plus some remarkable unphysical literalism display a lack of critical thinking. Yet some topics are done quite well.
- *Despagnat* is a survey by one of the more thoughtful lifetime advocates that quantum reality can't be reconciled with concrete realism in any form. It has good sections comparing different postulations that are not all consistent or independent.
- *Bell* collects his papers in *Speakable and Unspeakable in Quantum Mechanics*. Bell was simultaneously critical of the theory and a little over-involved. Some of the papers are worth reading to find out what is not there.
- *Bialnicky-Birula, Theory of Quanta* is a novel volume from an original thinker. Although conventional in layout there's not enough material to make a course. At the same time a number of interesting side topics are done imaginatively.
- *Van der Waerden* is a nice old congenial book with real educational value and solid thinking.
- *Feynman, Lectures III* was reported to be Feynman's goal in writing Volumes I and II. Feynman's magic fails when he gets tangled up trying to derive quantum mechanics from Stern–Gerlach and ad-hoc notation. The other two volumes on the rest of physics are priceless.

- *Mahan, Quantum Mechanics in a Nutshell* is inspired by *Davydov*, and both are very competent and well-organized. It avoids brainwashing preliminaries and postulates by simply not discussing them. While intended for the graduate level, it is about solving the Schrödinger equation and developing approximations, and falls short of developing transformation theory and much of the theory of angular momentum.
- *Schwable* has at least two books spanning the level from introduction to graduate topics. The math is skillful and not arranged to disable the student. The presentation puts everything as cut and dried with no critical thinking involved.
- *Greiner* and collaborators produced 13 or more books on almost every topic in theoretical physics. These books repeat what's found in other books while filling in the algebraic steps other books omit. Some of that is useful. It is disappointing to see how often they cite other Greiner books, and few of the actual sources.
- *Feynman and Hibbs*' book on path integrals is accessible to undergrads and novice grad students. The physical reasoning is delightful. The dust jacket reveals Feynman thought ordinary approaches to quantum mechanics were deeper than his own. This got omitted from the book, however.
- *Holstein* is a collection of relativistic quantum mechanics applications above the level of *Sakurai*, and well-suited to a second semester graduate course. Many topics have masterful treatments not found anywhere else.
- *Sakurai, Advanced Quantum Mechanics*, the old version circa 1965, is a wonderful transition book on relativistic quantum theory leading towards quantum field theory. Other books of that bygone era such as *Roman* have the simple good old stuff that modern books forget to include.

References

[1] Taylor G I 1909 *Proc. Cam. Philos. Soc.* **15** 114
[2] Darwin C G 1927 *Nature* **119** 282–4
[3] von Neumann J 1932 *The Mathematical Foundations of Quantum Mechanics* (Berlin: Springer)
[4] Darwin C G 1929 *Proc. R. Soc.* A **124** 375
[5] Mott N F 1929 *Proc. R. Soc.* A **126** 79
[6] Bell J S 1964 *Physics* **1** 195–200
[7] Bricmont J 2017 *Int. J. Quantum Found.* **3** 31 We put a sentence from the abstract at the end of the passage.
[8] Fine A 1996 *The Shaky Game: Einstein, Realism, and the Quantum Theory* (Chicago, IL: University of Chicago Press)
[9] Shi Y 2000 *Ann. Phys.* **9** 637
[10] Zeh H D 1996 *Decoherence and the Appearance of a Classical World in Quantum Theory* ed Joos Erich *et al* (New York: Springer)
[11] Powers T 2000 *Heisenberg's War: The Secret History Of The German Bomb* (Boston, MA: Da Capo Press)
[12] Einstein A, Podolsky B and Rosen B 1935 *Phys. Rev.* **47** 777

[13] Schrödinger E 1936 *Naturwiss.* **23** 807
[14] Hardy L 1993 *Phys. Rev. Lett.* **71** 1665
[15] Pittmann T B 1995 *et al Phys. Rev.* A **52** R3429
[16] Bennink R S *et al* 2002 *Phys. Rev. Lett.* **89** 113601

www.ingramcontent.com/pod-product-compliance
Lightning Source LLC
Chambersburg PA
CBHW061416210326
41598CB00035B/6236
* 9 7 8 1 6 8 1 7 4 1 6 2 8 *